D0982462

# The Birder's Bug Book

# The Birder's Bug Book

Gilbert Waldbauer

HARVARD UNIVERSITY PRESS

Cambridge, Massachusetts, and London, England

1998

Printed in the United States of America

Designed by Marianne Perlak

*Library of Congress Cataloging-in-Publication Data*

Waldbauer, Gilbert.
The birder's bug book / Gilbert Waldbauer.
p. cm.
Includes bibliographical references and index.
ISBN 0-674-07461-0 (alk. paper)
1. Birds—Ecology. 2. Insects—Ecology. I. Title.
QL673.W29 1998
595.717'8—dc21 98-3428

*To my beloved wife,*
*Stephanie Stiefel Waldbauer,*
*friend, valued adviser,*
*and discerning critic of my work*

# Contents

# Preface

In 1944 I took Aretas A. Saunders's beginning biology course at Central High school in Bridgeport, Connecticut. I didn't realize it at the time, but he and his two biology courses (I took the advanced one the next year) were setting the future course of my life. Mr. Saunders was an accomplished naturalist and ornithologist, a leader of the movement to identify birds through binoculars rather than over the sights of a shotgun, an acknowledged authority on the calls and songs of birds, and author of the 1935 *Guide to Bird Songs,* a book that was *the* authoritative source before sound recordings came into wide use. On weekends he took me and my friend Robert Braun on birding expeditions to nearby areas. (Long trips were out of the question because of wartime gasoline rationing.) We visited Connecticut "hot spots" such as the Wood Duck Woods in Fairfield, the salt marshes at Westport, and the tidal flats near Norwalk.

Since my mother couldn't afford to buy me a pair of binoculars, I squinted at birds through a three-power "monocular," actually half of a pair of ancient field glasses that I had bought for fifty cents in a pawn shop. On one trip to the marshes at Westport, we were accompanied by Elting Arnold, who had come from Washington, D.C., to bird with Mr. Saunders. That day I saw my first yellow-crowned night heron, then a rare bird in Connecticut. Mr. Arnold noticed my "monocular" and apparently took pity on me, for a few weeks later Mr. Saunders passed on to me a package from Elting Arnold that contained a used but still very serviceable pair of 8 × 30 binoculars.

It wasn't long before I decided that I wanted to be a biologist, preferably an ornithologist. At the senior class graduation banquet, one of the students read aloud the "class will," which was actually a prediction of what would eventually become of the members of the class of 1946. The will was,

at least in my case, amazingly prophetic. The reader intoned (the words are burned in my brain): "Gilbert Waldbauer, Professor of Entomology at Yale University, leaves a well-mounted tarantula to anyone who will have it." It seemed funny then, but I was a bit miffed. I wanted to be an ornithologist—not an entomologist and certainly not a professor. At that time I imagined, quite incorrectly, that professors spend all their days in classrooms and are seldom outdoors, where natural history actually happens.

But the class will was right and I was wrong. I did an undergraduate major in entomology at the University of Massachusetts in Amherst, but not until after a hitch in the U.S. Army that earned me the GI Bill credit that supported me as an undergraduate. I took the one ornithology course that was offered at UMass and thoroughly enjoyed learning the scientific foundation of the field. In the meantime, I had met Charles P. Alexander, head of the Department of Entomology and an outstanding researcher in the field of insect taxonomy. Professor Alexander was vibrant, very much interested in students, a knowledgeable all-round naturalist, and an exuberantly enthusiastic student of the insects. When I realized that he was willing to take me under his wing, I became his grateful protégé.

I soon began to think about a career in entomology. Insects were as fascinating as birds, and I envisioned a career devoted to research on the interrelationships between birds and insects, a hope that was ultimately realized. My thought eventually became a decision. I would become an entomologist. Professor Alexander assured me that my grades were good enough to qualify me for a teaching or research assistantship that would support me while I was a graduate student. He also told me that jobs were scarce for ornithologists but more plentiful for entomologists. After graduating from UMass, I entered the Department of Entomology at the University of Illinois in Urbana-Champaign as a graduate student and teaching assistant. I have been there ever since.

Off and on during my professorial career, I thought about writing a book on the many complex and interesting interrelationships between insects and birds. I envisioned a scholarly volume aimed at professional ecologists and evolutionary biologists and written in the inscrutable language of biology. I never wrote that book. When I retired from teaching, I decided that I would try to become a liaison between professional biologists and birders, other amateur naturalists, and anyone else interested in the interactions of animals. Thus I came to write *The Birder's Bug Book,* scientifically accurate but couched in language that avoids the professional jargon of biology insofar as possible.

But why write this book? The answer is that a bird—or any other organism—is best and most instructively viewed in the light of its ecological context, the plants, insects, and other organisms with which it associates. And as you will discover, insects are a significant and often hugely important part of the ecological context of most birds—as are birds an important part of the ecological context of insects. Trying to understand a bird or an insect taken out of its ecological setting is as unrewarding and futile as trying to see significance in a single word lifted from its context on this page.

This book is intended for anyone who is interested in natural history, especially for those who are interested in insects or birds. In my experience, almost all birders—even the most dedicated listers—view a bird as more than a checkmark on a list. They are curious about its life history and behavior, both of which may be largely determined by the insects that it eats and the insects that eat it. These birders, especially those interested in the conservation of bird species, want to know about the ecological context in which birds live. They know that birding becomes more and more enjoyable as we learn more about the birds that we see, and they also know that we cannot hope to save a species from extinction in the wild unless we understand it as a functioning member of its ecosystem.

The Birder's Bug Book

# Bugs and Birds through the Ages

# 1

The tap-tap-tap of a downy woodpecker sounds crisply through the clear, cold air of a snowy woodland in New Hampshire. Late in summer, a house sparrow, feathers puffed out and partly folded wings fluttering, takes a dust bath in a small, dry depression beside a well-worn footpath in an Illinois farmyard, and in Tennessee an orchard oriole bends its head back as it rubs an ant against its flight feathers. On a prairie pond in Saskatchewan, a Wilson's phalarope spins around in tight circles as it swims in shallow water. In Africa, a small, sparrowlike bird approaches a honeybadger and entices it to follow, flitting from perch to perch as it gives distinctive calls. A starling in Michigan adds a sprig of fresh green leaves to its nest in an abandoned woodpecker hole. In Colorado an American dipper clambers over rocks beneath the surface of a fast-flowing stream, and in Connecticut an eastern towhee kicks noisily as it rummages among dead leaves in a woodland.

These seemingly unrelated activities have something in common: all stem from an association between birds and insects. The woodpecker eats the insects that it uncovers by chiseling through the bark of a tree. A dust bath may kill some of the lice that live on the house sparrow, and lice and mites that live among the oriole's feathers will be killed by formic acid or other toxic substances produced by the ant. The phalarope's gyrations stir up the bottom sediments and bring to the surface the aquatic organisms, including insects, on which it preys. The African bird, known as the greater honeyguide, leads the honeybadger, or even a person, to a colony of honey bees in a hollow tree and later eats the scraps of beeswax and honey that remain after its willing helper has broken open the tree to get at the honey. The fresh green sprigs that the starling adds to its nest contain chemicals that deter or kill parasitic mites. The dipper searches for a meal of aquatic

insects that cling to rocks in the stream, and the towhee eagerly snaps up the insects that it exposes by kicking aside dead leaves.

The relationships between birds and insects are ancient, and these often highly specialized interactions have become prominent and important facets of almost all continental ecosystems. Just try to imagine a woodland without its tanagers, flycatchers, vireos, warblers, woodpeckers, thrushes, nuthatches, and many other insectivorous or partially insectivorous birds. A woodland without insects would be bereft of these marvelous birds. There would be no swallows, swifts, or nighthawks wheeling in the sky if there were no "aerial plankton" for them to eat, the flying insects that they scoop up as they swoop through the air.

Through their specific associations, birds and insects have had significant and far-reaching effects on each other's evolutionary paths. The anatomy and behavior of many birds have been specially modified for the taking of their particular insect prey. Witness, for example, the flight of the whip-poor-will as its gaping mouth, widened by a fringe of bristles, traps insects from the night air; the antics of a red-billed oxpecker as it plucks ticks from the skin of an African rhinoceros; the nervous movements of a wood warbler as it uses its tweezerlike bill to pick small caterpillars and other insects from the leaves of a tree; and the darting flights of an olive-sided flycatcher as it sallies forth to use its snap-trap bill to intercept flying insects that it spied from its perch at the top of a balsam fir. Insects have, in turn, responded to the depredations of birds by evolving a variety of defensive tactics, ranging from exquisite camouflage and swift flight to the venomous sting of a bee or the vomit-inducing properties of a monarch butterfly. Most venomous or toxic insects have bright, easy-to-remember color patterns that birds and other insectivores can readily learn to avoid, a fact that makes it possible for some harmless insects which lack defenses of any sort to escape predators by bluffing, by mimicking the color patterns and even the behavior of insects that do have defensive weapons.

The wood-probing, insect-eating huia of New Zealand is an avian example of extreme anatomical and behavioral specialization for the capturing of insect prey. The huia, last seen in 1907 and now extinct, was unique among the birds of the world, because of the radically different bills of the two sexes. The straight, stout bill of the male was less than half the length of the slender, decurved bill of the female. According to the New Zealand

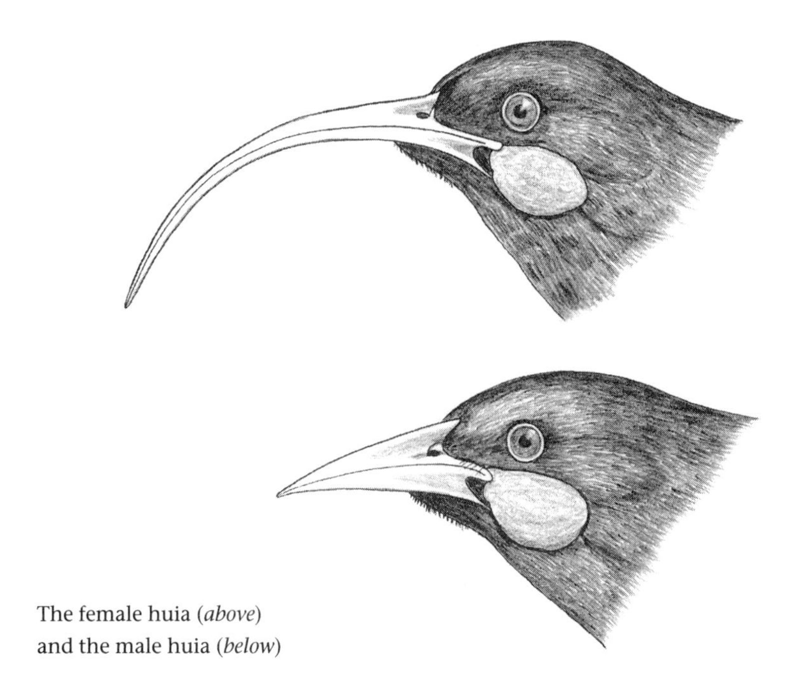

The female huia (*above*) and the male huia (*below*)

ornithologist Walter Buller (see E. G. Turbott's annotated edition of Buller's 1883 book), foraging by huias was a cooperative effort between the male and the female, with their different bills complementing each other. Buller described how a captive pair, which seemed to be inseparable, attacked a decaying log to obtain their favorite food, larvae of the huhu, a very large, wood-boring, long-horned beetle. Having noted that their behavior seemed to show the "usefulness of the differently formed bills of the two sexes in the economy of nature," he went on to describe how the huias responded to a rotting log that he placed in their cage:

> They at once attacked it, carefully probing the softer parts with their bills, and then vigorously assailing them, scooping out the decayed wood till the larva or pupa was visible, when it was carefully drawn from its cell . . . and then swallowed. The very different development of the mandibles in the two sexes enabled them to perform separate offices. The male always attacked the more decayed portions of the wood, chiselling out his prey after the manner of some Woodpeckers, while the female probed with her long pliant bill the other cells, where the hardness of the surrounding

> parts resisted the chisel of her mate. Sometimes I observed the male remove the decayed portion without being able to reach the grub, when the female would at once come to his aid . . . I noticed, however, that the female always appropriated to her own use the morsels thus obtained.

Not only do birds exploit insects, but insects exploit birds. The arrival of birds on the evolutionary scene provided a new resource for insects. Nature abhors a vacuum, including ecological vacuums, and insects soon began to take advantage of this new resource. Modern insects have, through the evolutionary process, come to exploit birds in many ways. Some, such as certain kinds of mosquitoes, visit birds from time to time to suck blood, the source of the protein needed to develop their eggs. Other insects and many mites, tiny, eight-legged relatives of the insects, live in the nests of birds—some acting as scavengers of dead organic material and others attacking the nestlings or even the adults. Some fleas live in the nests of birds throughout their lives. The legless larvae eat organic debris, and the wingless adults, which lay their eggs in the nest and may spend much of their time there, use their jumping legs to leap onto the body of the resident bird when they require a blood meal. Some insects, such as the biting lice, have lost their wings and live permanently as parasites on the bodies of birds. They usually die when the host bird dies and can transfer from one bird to another only when birds are in close bodily contact. Some of these lice are generalists that can live on many different kinds of birds. Others have become so specialized that they can live only on one species of bird or on a few closely related species. If a specialist louse is artificially transferred to a bird other than its usual host, it will refuse to eat and soon will die.

✱ It seems inevitable that many birds have come to depend upon insects as food and that some insects have evolved to exploit birds as food. After all, insects occur, usually in large numbers, almost everywhere that birds occur—all over the world except on the Arctic and Antarctic icecaps. But they are surprisingly abundant in the parts of the Arctic that are not permanently covered by ice, and some occur even in the few wind-blown, ice-free areas of the Antarctic. They are very rare in the seas, but mosquitoes breed in brackish seashore pools, and marine water striders skate on the surface of the ocean hundreds of miles from shore.

The diversity of insects is mind-boggling. Of the approximately 1.2 million animal species now known, about 900,000—75 percent—are insects, and there may be several million insects that have yet to be discovered. The beetles alone number 300,000 known species, and are thus more than five times as numerous as all of the vertebrates put together—the fish, amphibians, reptiles, birds, and mammals.

Insects are also numerous as individuals. In Pennsylvania, a square foot of forest leaf litter and humus examined in the 1940s contained almost 10,000 arthropods, including nearly 7,000 mites, almost 3,000 insects, and a few other arthropods such as centipedes and millipedes. The insects on an African savanna outweigh the large grazing animals on a per-acre basis, and in an abandoned, weed-grown crop field in North Carolina surveyed in the 1960s, the plant-feeding insects alone outweighed all the sparrows and mice by a factor of nine. A single honey bee colony may contain more than 50,000 workers, and a tropical termite colony may include well over 1 million individuals.

Insects derive a living from their many habitats in almost every imaginable way. In the aggregate, they feed on almost all of the 250,000 known plant species. Some serve the plants as pollinators, and others consume leaves, fruits, roots, or even the woody stems. Other insects eat carrion or dead vegetation. Many attack insects and other animals: some as blood feeders; some as predators, especially of other insects; and a sizable group as parasites of earthworms, other insects, birds, mammals, and other creatures.

Because of their diversity and abundance, insects are important and usually indispensable parts of almost all the terrestrial and freshwater ecosystems on earth. Consider only insects that pollinate plants. Without them, tens of thousands of plants would become extinct or survive only as remnant populations. The land would be dominated by pines, firs, grasses, and a few other wind-pollinated plants such as cottonwoods and ragweeds. If we consider only sweet fruits that people eat, without pollinating insects there would be no melons, figs, peaches, plums, apricots, cherries, strawberries, raspberries, blackberries, blueberries, cranberries, kiwis, citrus fruits,

pears, or apples. And if we consider that insects perform many essential functions other than pollination, it becomes apparent that without insects most of the ecosystems of the continents would collapse and be replaced by degraded ecosystems that would be far less hospitable to humans, birds, and most of the other forms of life that have thus far evolved.

The insects are the greatest evolutionary success story of all time. Millions of years before the first bird evolved, they had already spread throughout all the continents. The first primitive insects, wingless, crawling creatures, appear in the fossil record about 400 million years ago during the age of fishes, when the earliest amphibians were venturing onto the land—long before the first reptile or mammal had evolved, and even longer before the first bird appeared in the fossil record. Winged insects first appear 65 or 70 million years later, about 330 million years ago during the Carboniferous Period, the heyday of the great coal forests, the age of the amphibians, and the time of the appearance of the very earliest reptiles. By about 250 million years ago, when the ancestral Appalachian Mountains were being formed, when the first dinosaurs appeared, and long before the birds evolved, the insects were already an ancient group, many of whose major subdivisions (orders) were already present, including quite a few that are familiar to us today, such as dragonflies, crickets, true bugs (the members of the order Hemiptera), and beetles. The wasps, bees, and butterflies appeared much later.

Insects are relatively rare in the fossil record, less numerous than the bones of vertebrates, the shells of mollusks, or even the soft bodies of some marine invertebrates. Most insect fossils are fragmentary, often no more than a wing, but a few—especially some in amber or fine-grained shales—are complete and surprisingly well preserved. There were giant insects during the Carboniferous Period: dragonflies and mayflies with wingspans of 28 and 18 inches, respectively, and a formidable piercing and sucking species that belonged to a now extinct order and had a stout, inch-long beak and a 22-inch wingspan. The best preserved insect fossils are embedded in amber, the fossilized resin of pines and other trees. Many are in astonishingly good condition—every bristle in place and looking as if they had died just yesterday. The Baltic amber of Europe, used in jewelry for thousands of years, is justly famous for its insect inclusions, but amber containing insects is also found in a score of other sites in Asia, Europe, North

*Mischoptera,* a fossilized insect that lived about 300 million years ago

America, and the Caribbean. The oldest known ant, contemporaneous with dinosaurs, was preserved 100 million years ago in a lump of amber found in New Jersey in 1966. Insects in Baltic amber, only about 40 million years old, look familiar and are very similiar to modern species, but older amber, such as that from New Jersey, contains insects that are obviously different from modern species.

As you already know, birds evolved much more recently than the insects. The first animal with undisputed feathers does not appear in the fossil record until about 150 million years ago, during the Jurassic Period, when huge dinosaurs roamed the earth, when conifers and palmlike cycads were the dominant land plants, and before the angiosperms, or flowering plants, had evolved. This crow-sized creature had many small, sharp teeth, feath-

ered wings, and a long tail with the feathers arranged along its sides as the "leaflets" of a fern are arranged along the stem of the frond. In 1862, this creature was named *Archaeopteryx* (from the Greek roots *archeo,* ancient, and *pteron,* wing) by the German scientist Hermann von Meyer. Some modern authorities consider *Archaeopteryx* to be a bird while others think of it as a feathered reptile, but most agree that it is an ancestor or a close relative of the ancestor of modern birds. (Although there were once pterosaurs, flying reptiles, they were certainly not the ancestors of the birds. Their wings, rather than being feathered, were membranous like those of a bat.)

The fossils of *Archaeopteryx* were found in Bavarian quarries from which exceedingly fine-grained Jurassic limestone is mined for use as lithographic slabs. The first one found is an impression of a single feather, but since then six others have been discovered. Some of these fossils are fragmentary, but one that is in the Humboldt Museum für Naturkunde in Berlin is beautifully preserved. As the illustration on the facing page shows, there are unmistakable imprints of feathers, and the skeleton is virtually complete and fully articulated.

There are two theories of the evolutionary origin of the birds. One is that they descended directly from early reptiles, the thecodonts, which also gave rise to the crocodilians and the dinosaurs. This theory places the origin of the birds about 230 million years ago in the middle of the Triassic Period. The opposing and most generally accepted theory is that the birds did not stem directly from the thecodonts, but that they arose much more recently, somewhat more than 150 million years ago, during the Jurassic Period, from dinosaurs that descended from the thecodonts. According to the latter theory, the birds began with *Archaeopteryx* or some similar creature. The theory of the Triassic origin of birds leaves a difficult-to-explain gap of 90 million years between the presumed origin of the birds and the appearance of *Archaeopteryx.* Some think that a fossil recently found in Texas closes this gap, since it is 75 million years older than *Archaeopteryx.* Originally considered to be fragments of a bird, it was named *Protavis,* but many authorities now think that the condition of this fossil is so poor that it cannot be shown to represent a bird. It is fragmentary, includes no feather imprints, and when found consisted of a jumble of broken and unarticulated bones. There is no doubt that birds descended from reptiles, but whether they arose from dinosaurs or earlier reptiles is being hotly debated. So far there is

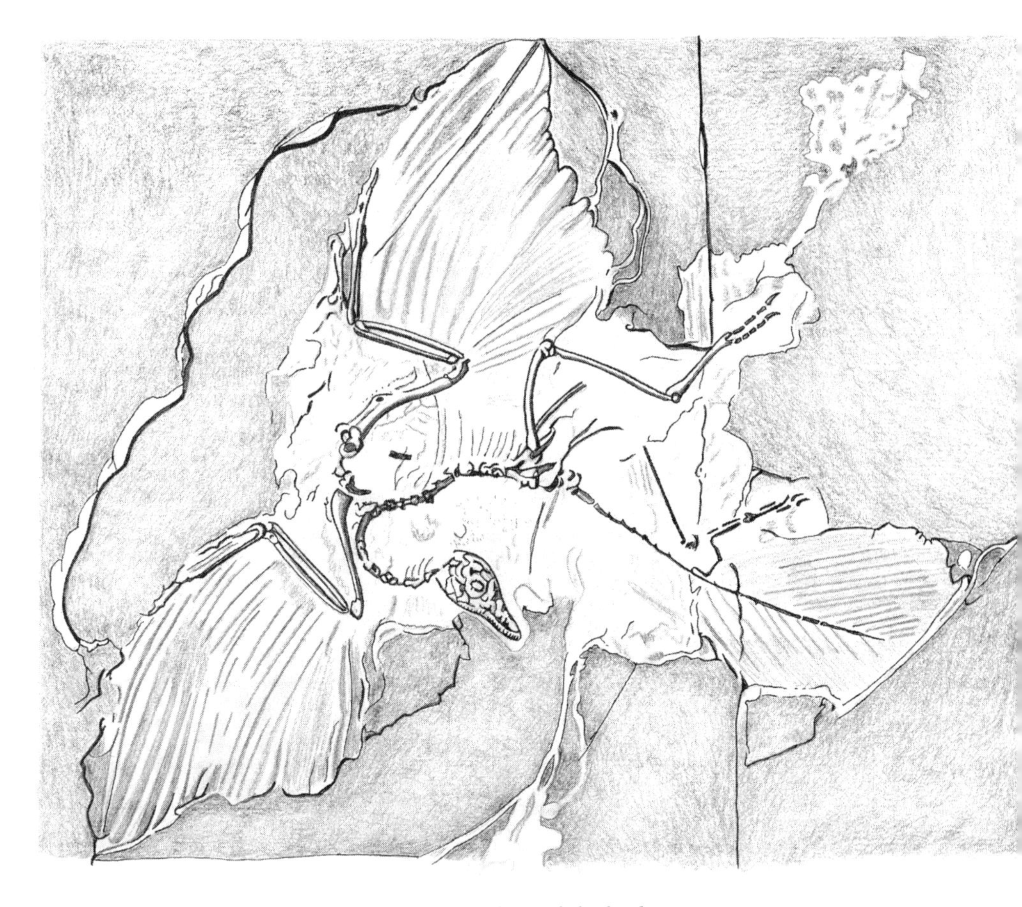

*Archaeopteryx,* a "missing link" between the reptiles and the birds

no conclusive fossil evidence for either one of these theories. In a recent book, Alan Feduccia of the University of North Carolina argued against the dinosaur origin of birds. But a fossil just discovered in China, an obvious dinosaur adorned with what appear to be feathers, supports this theory (see the news report in the November 1, 1996, issue of *Science.*) In either case, whether the birds appeared about 150 million years ago or as much as 230 million years ago, insects—both wingless and winged species—were already there to greet them.

According to recent estimates, about 9,600 species of birds are known to science, and ornithologists are agreed that very few unknown species re-

main to be discovered. The birds are outnumbered by just the *known* species of insects by better than 80 to 1 and by the beetles alone by more than 30 to 1. Although the birds are few in number of species, they have, like the insects, invaded virtually all of the terrestrial and freshwater habitats, and they have been far more successful than the insects in conquering marine habitats.

The numerous and complex associations between birds and insects that we see today had their beginnings even before the birds evolved—with insect-eating reptilian ancestors of the birds. The insects, already numerous in species and probably abundant as individuals, were a food resource that the early reptiles must have exploited and that modern reptiles continue to exploit. Some of the predaceous reptiles, including some of the dinosaurs, were much too large to subsist on creatures as small as insects. But from the very beginning of the reptilian line, and even among the dinosaurs, there have been small, agile species, sometimes growing no larger than pigeons, that probably subsisted mainly by eating small vertebrates and insects. Furthermore, the newly hatched young of the egg-laying dinosaurs may have been too small to eat creatures much larger than insects, just as are the hatchlings of our modern alligators and crocodiles. It may well be that as hatchlings some of the large carnivorous species of dinosaurs, perhaps even the mighty *Tyrannosaurus rex,* fed on insects.

No one really knows what *Archaeopteryx* ate, but its small sharp teeth, all more or less alike, are reminiscent of the teeth of some of our modern insectivorous lizards and snakes. Thus it is not unlikely that this lizardlike bird included insects and other small creatures in its diet, as do many modern-day birds. The earliest insectivorous birds, including *Archaeopteryx,* were probably omnivorous generalists rather than specialists. But from these generalists, opportunists that ate whatever they could find, evolved the many highly specialized insect-eating birds that we see today.

As the late Herbert H. Ross pointed out in his textbook of entomology, the rapid evolution of the amphibians during the Carboniferous Period, which began about 350 million years ago, was probably made possible by the new and abundant food supply that consisted of the recently evolved freshwater arthropods, particularly the aquatic nymphs of insects. During the Tertiary Period, which began about 65 million years ago, roughly 300 million years after the beginning of the Carboniferous, there was a similar rapid evolution among the birds, especially the passerines, or perching birds (also known as the songbirds)—probably in response to the new food resources provided by the rapid proliferation of insects that was, in turn, a response to the appearance and burgeoning of a new and more varied plant resource for insects, the angiosperms. Today most insects depend for their sustenance either directly or indirectly upon these flowering plants. Their leaves, stems, flowers, pollen, fruit, and roots are all fare for insects, and they are indirect support for predaceous and parasitic insects that attack the plant feeders and for scavenging species that eat the remains of dead plants and dead insects.

Compared to insects and birds, people are newcomers who arrived on the world scene only yesterday. The first humans, members of the genus *Homo,* appeared about 2 million years ago, but our own species, *Homo sapiens,* did not appear until about 250,000 years ago. Ever since then, people have interacted with insects and birds, an interaction whose nature changed as human culture evolved from hunting and gathering to the complex technology of today.

The few hunter-gatherers that survive today regularly include insects in their diets. Indeed, as so interestingly told by Friedrich Bodenheimer, insects—especially grasshoppers, termites, beetle grubs, and caterpillars—are eaten by people all over the world except for most members of Western cultures. Other insects are a detriment to people. Some of them have always bitten, infested, or otherwise attacked our bodies. With the development of agriculture between 8,000 and 10,000 years ago, insects that attack crop plants or domestic animals became a serious threat to our major source of food.

Birds always have been and continue to be exploited by people everywhere for their flesh, feathers, eggs, and in the case of certain seabirds, even their excrement—the guano that is used as fertilizer. Some primitive societies, probably as a consequence of their exploitive interest in birds, are impressively well acquainted with them. According to Jared Diamond, the Fore people of New Guinea have 110 different names for the 120 species of

birds, other than cassowaries, that occur in their area. With only a few exceptions, the Fore *ámana áke* (species) correspond to species recognized by ornithologists.

Few societies have comparably detailed vernacular classifications of insects. Even the Fore have not named the large and spectacular butterflies that are all around them. Our own language has few "folk" names for insects, although entomologists have created English names for pest insects. The Japanese, however, have always had a special fondness for dragonflies and knew most of the 200 species found on their islands by vernacular names long before the advent of entomology in their country.

North American ornithologists of the late nineteenth and the early twentieth centuries categorized birds according to their economic impact on humans. A bird was judged to be economically harmful if it damages crops or other resources valued by people, or economically beneficial if it destroys weed seeds, harmful insects, or other organisms inimical to human interests. Economic ornithologists, notably W. L. McAtee and F. E. L. Beal of the Biological Survey of the U.S. Department of Agriculture, killed thousands upon thousands of birds in order to examine the contents of their stomachs. From their findings, they concluded that few birds are harmful, and that most of them are, from an economic perspective, either neutral or useful, usually because they eat destructive insects. After all, any damage done by a bird may, on balance, be outweighed or equaled by some good that it does. For example, although meadowlarks may occasionally eat corn seeds, this trifling harm is greatly outweighed by the good they do by eating weed seeds and insects destructive to corn. The data accumulated by the early economic ornithologists are still useful, and similar studies are still occasionally necessary. But the killing of wild birds for research purposes is now controlled, and their wholesale destruction by scientific investigators is largely a thing of the past.

We no longer view birds as if through the eyes of a cost accountant, valuing them only if they make an immediate contribution to our economy. We appreciate them for their beauty, and see them as integral and essential elements in the ecosystems of our planet. Esthetically, we enjoy their grace, the loveliness of their plumage, and the vibrance of their songs. We know that the meadowlark's song is worth more than money. Ecologically, we realize that birds play many different, complex, and necessary roles in ecosystems, interacting with other organisms, especially insects, in thousands of different ways. Take, for example, the hummingbirds of the Americas. These 319 species coevolved with the plants that they visit for nectar. The plants accommodate the hummingbirds by the large size of their blossoms, their abundant nectar, and their red color, the favorite hue of hummingbirds. The hummingbirds reciprocate by pollinating the plants, many of which could not exist without them. In turn, these plants are home and sustenance for many different kinds of herbivorous insects and, indirectly, for the parasites and predators of the herbivores. All of these insects are, in their turn, food for birds, including hummingbirds.

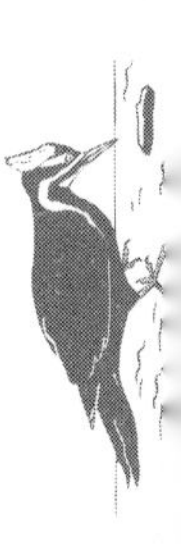

# The Only Flying Invertebrates

# 2

All of us are aware—sometimes all too aware—that insects can fly. As we try to fall asleep at night we are sometimes disturbed by the high-pitched hum of the wings of a female mosquito as she cruises in search of a blood meal that will most likely come from one of us. In the summer of 1996, as I stood quietly in the forest at Algonquin Provincial Park, Ontario, waiting for a black-backed woodpecker to return to its nest hole, black flies swarmed around me, waiting for the opportunity to suck my blood by snipping through my skin with their sharp, scissorlike mandibles. But this bird was a lifer for me, a species that I had never seen before, a potential addition to my life list of birds. I was not about to follow the example of the moose that had retreated to clear areas along the roads to escape the pestiferous flies.

Our experiences with winged insects are not always so unpleasant. As we walk along the edge of a pond, we hear the rustle of a dragonfly, membranous wings glistening in the sun as it swerves past us in fast pursuit of its insect prey. A monarch butterfly slowly waves its beautiful orange and black wings as it sips nectar from goldenrod flowers. Above the canopy of a rain forest in Mexico, a giant morpho butterfly flaps along, its iridescent blue wings reflecting the sun as brightly as a heliograph. All of us have been heartened by the cheerful chirps of crickets, charmingly described by the late Vincent Dethier in *Crickets and Katydids, Concerts and Solos:*

> The instrument of this indefatigable musician is his interlocking front wings, the stiff parchment-like covers for his membranous hind wings. As the bow of a violin is drawn across the strings and sets them vibrating and as the body of the violin is set resonating by transmission of the vibrations through the bridge, so the cricket draws a scraper across a file of small teeth and sets the wing covers to resonating. The wing covers are raised in song at an angle of about forty-five degrees and brought to-

gether periodically. With this closing motion, a ridge on the upper wing cover scrapes across a file on the lower wing cover and generates a high-pitched sound.

Insects are the only invertebrates, animals without backbones, that ever evolved wings and the power of flight. If you see an animal that does not have a backbone but that does have wings, you know with certainty that it is an insect. But you cannot say the opposite. An invertebrate that lacks wings may, nevertheless, be an insect. A few insects are wingless throughout their lives, and the winged species, except for some that have externally developing wing buds, have no visible wings during their growing stage.

Among the vertebrates, the animals with backbones, only three groups, the birds, the mammals, and the reptiles, ever evolved the power of true flight. The birds and the bats are the only surviving vertebrates with wings, but millions of years ago, during the age of reptiles, there lived the flying reptiles, or pterosaurs, that I have already mentioned. The modern vertebrates other than birds or bats that are said to fly are capable only of gliding, not of true flight. The flying squirrel glides on flaps of furry skin that stretch between its front and hind legs. The webbed and greatly broadened feet of the flying frog of Java serve as gliding surfaces, and the flying fish of tropical waters merely glides on its expanded pectoral fins.

The wings of birds, bats, and the erstwhile pterosaurs are all modified front limbs that were originally used for walking by their earthbound ancestors. Not so the wings of insects. They are definitely not modified legs; they originated as flattened outgrowths of the body wall that probably enabled the ancestors of winged insects to glide. The two main theories of the origin of insect wings are well summarized by Howard E. Evans. The "flying fish" theory postulates an aquatic ancestor that climbed out of the water onto vegetation—perhaps to feed or to escape enemies—and used broadened gill plates to glide back to the water's surface. According to the alternative "flying squirrel" theory, an arboreal ancestor developed flanges on its thorax that enabled it to glide from tree to tree or down to the ground. But both theories agree that it all began with gliding forms that eventually developed the wing hinges and muscles required for powered flight. The "flying squirrel" theory, tree climbing followed by gliding, is a popular hypothesis for the origin of flight in birds.

The ability to fly is so important to most adult insects that we cannot

even begin to understand their lives unless we understand how they manage to conquer the air. Like the body of a bird, the body of an adult winged insect is a finely integrated flying machine, its structure largely dictated by the need to fly and all its organ systems dedicated in varying degrees to serving this function. The "skeletal" system is light in weight to ease the burden on the wings; the sensory and central nervous systems guide and control flight; the organs of ingestion and the digestive system obtain and process high-energy food that provides the fuel that will be carried to the flight muscles by the circulatory system; the respiratory system provides the large quantity of oxygen required by the flight muscles to metabolize the fuel that powers them; and even the endocrine system is involved, secreting hormones that determine at what season and for how long some insects will fly.

An insect's body is anatomically and physiologically so different from our own, or from that of a bird or any other vertebrate, that we must have some comprehension of it if we are to understand either winged or wingless insects. All insects, even the tiniest gnats, have inner organ systems that are comparable to our own in complexity and that are competent to support life styles and behaviors that are often intricate and surprisingly sophisticated. There are, however, major differences between insects and vertebrates such as birds in the structure and function of these internal systems.

The circulatory and respiratory systems are a case in point. In humans, birds, and all other vertebrates, these two systems are intimately related and cooperate to supply the cells of the body with the oxygen that they require. In birds, as in all vertebrates, the circulatory system is closed, with the blood confined in a "plumbing system" of arteries, capillaries, and veins. The blood is oxygenated in the lungs, and the heart then pumps it through the arteries and capillaries so that every cell of the body is supplied with oxygen that is carried by the hemoglobin in the red cells of the blood. The blood also distributes hormones throughout the body, transports nutrients from the digestive system to the cells, and picks up the waste products of metabolism from the cells. The capillaries then come together like the tributaries of a river to form the veins that carry the blood back to the heart via the kidneys, where nitrogenous and mineral waste products are removed. Carbon dioxide, the gaseous waste product, is eliminated by the lungs as the blood is reoxygenated.

In insects, by contrast, the circulatory and respiratory systems are independent of each other. The circulatory system does not carry oxygen to the cells, and the blood of insects is generally not red because there is no need for the red, oxygen-carrying hemoglobin. Insects have open circulatory systems; the blood flows freely in the body cavity, which extends out into the appendages, rather than coursing through a closed system of "pipes." The insects' equivalent of a heart is an open-ended tube that pumps blood from the abdomen toward the head, thus creating a circulation within the body cavity that carries hormones and nutrients to the cells and soluble waste products from the cells to the Malpighian tubes, named for Marcello Malpighi, the seventeenth-century Italian anatomist who first discovered them. These slender tubes, the insectan equivalent of the kidneys, lie in the body cavity, absorb waste products from the blood, and empty them into the hind intestine.

Oxygen reaches the cells of the insect's body directly through a complex web of air-filled tubes, tracheas, that have access to the outside atmosphere by way of openings in the cuticle called spiracles. Oxygen diffuses through the spiracles and then along the tracheas, which branch and rebranch until each cell of the insect's body has its own contact with the atmosphere. As oxygen diffuses inward through a trachea, carbon dioxide diffuses outward. This simple diffusion supplies enough oxygen for some sedentary insects, but it does not supply enough to meet the needs of a rapidly flying insect. In such an insect, respiratory movements of the abdomen suck air into inflatable air sacs and then flush it through the tracheal system, thus increasing the oxygen supply.

The digestive systems of birds and insects vary in structure and physiology according to the foods that they eat. Both birds and insects include dietary generalists such as crows and cockroaches, omnivores that eat a variety of both animal and vegetable foods. There are also dietary specialists. Among the specialist birds are hawks that eat other birds, cuckoos that eat caterpillars, and crossbills that eat the seeds of conifers. Yellow-rumped warblers and tree swallows are even more oddly specialized. In a 1992 arti-

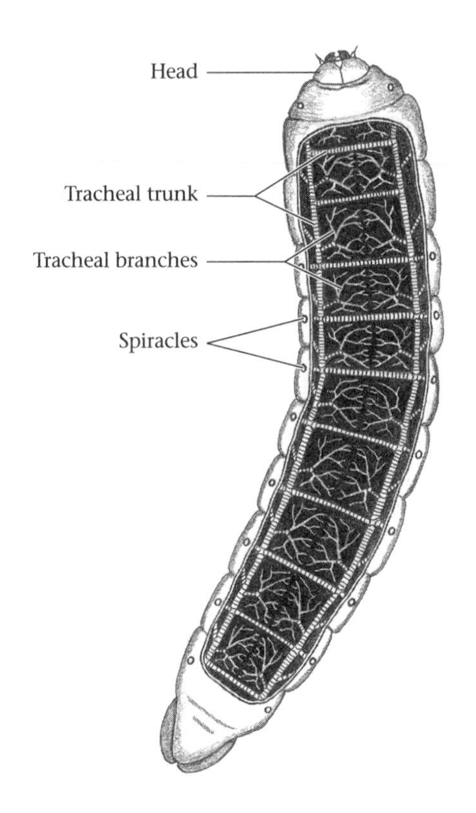


View from above of a caterpillar cut open to show the ladderlike tracheal system and some of its many branches

cle in the *Auk,* Allen Place and Edmund Stiles showed that the gut physiology of these birds is specifically designed to digest wax, which most animals cannot digest. In the summer both species feed mainly on insects, but during the winter they feed heavily on the wax-coated berries of bayberry and wax myrtle. Insects tend to be even more specialized. Some lice will live on only a few closely related species of birds; some parasitic insects will use as hosts only one or a few related insects; and many insects feed on only a few closely related plants and will starve rather than eat a plant of the "wrong" species.

The gut of an insect consists of three parts: the foregut, midgut, and hindgut. As in a bird, different areas of the insect's gut are specialized for different functions. The crop of the foregut is concerned mainly with the storage of food, the midgut with the production of digestive enzymes and the absorption of digested nutrients, and the hindgut with regulating salt balance and resorbing water from the fecal material. In some insects a muscular part of the foregut functions to grind food as does the muscular gizzard of a bird. In birds the grinding is done by grit or gravel that is swallowed. In insects, hard teeth that line the "gizzard" do the grinding.

The excretory systems of insects and birds perform similar tasks, although they evolved independently and differ in structure and physiology. The Malpighian tubes of insects, as you already know, absorb soluble wastes from the blood and empty them, the "urine," into the hind intestine, where they mix with the feces and are eliminated through the anus. Similarly, the kidneys of birds pass the urine into the lower (hind) intestine to be eliminated with the feces.

One of the most important functions of the excretory system of any animal is to get rid of nitrogen-containing wastes, the toxic byproducts of the metabolism of proteins. Most mammals eliminate them in the form of urea, which is toxic and soluble and must be flushed out of the body with large quantities of water. But many insects, reptiles, and birds have

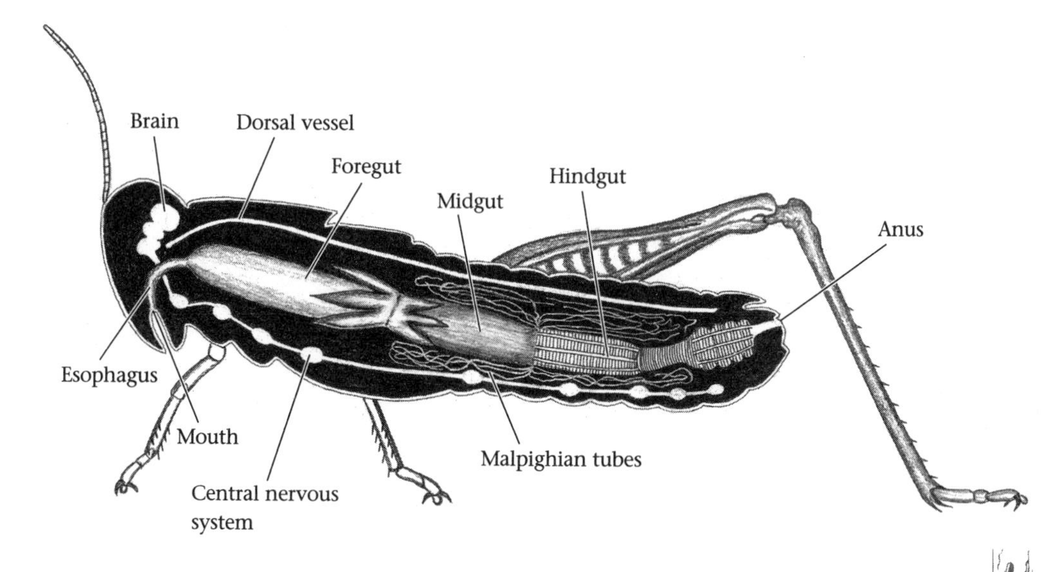


Side view of a grasshopper cut open to show some of its organ systems

little or no access to water other than that in their food and must conserve the water in their bodies. They cannot afford to lose the water needed to eliminate urea. Many of these animals eliminate nitrogenous wastes largely in the form of uric acid, which is less toxic and much less soluble than urea and can be eliminated from the body with very little water. Uric acid gives the droppings of birds their characteristic look, appearing as an extensive and almost solid white deposit on the surface of the dark food residue.

The central nervous system of a bird consists of a spinal cord and a relatively large and complex brain. As in other vertebrates, the cerebral hemispheres of the forebrain are the seat of intelligence and control sensory integration, complex instinctive behaviors, and learning. Simple behaviors can be controlled elsewhere in the brain. For example, a falcon with its cerebral hemispheres removed is still capable of catching a mouse, but having done so it just sits with the mouse in its talons, apparently not knowing what to do with it. Some birds are surprisingly intelligent. Ravens can be taught to count to seven; in England blue tits and great tits figured out how to remove or pierce the caps of milk bottles to get at the cream—a behavior that spread widely through the tit populations, apparently through observational learning; and, finally, several species of birds are known to use tools, a behavior that suggests insight learning. A green heron was observed using bits of bread as bait to attract fish within striking range of its beak. And the Egyptian vulture breaks open ostrich eggs by dropping large stones on them.

Some birds have amazingly good memories. Among them are certain species of chickadees, nuthatches, jays, crows, nutcrackers, and other birds that store seeds to be eaten when food is scarce. As Diana Tomback has observed, in late summer and fall, a Clark's nutcracker of our western mountains tucks away as many as 33,000 pinyon pine seeds in thousands of separate small caches in the soil or forest litter. As much as eight months later, the bird relocates its caches and recovers the seeds by poking its bill down into the soil. By tracking nutcrackers and noting whether or not a bird had discarded the hulls of seeds near a poked hole, Tomback determined that they were successful in relocating a cache at least 72 percent of the time. The true success rate was probably close to 100 percent, since apparent failures to relocate a cache were probably due to pilfering of the seeds by mice or the bird's removing the seeds and hulling and eating them elsewhere.

Like these unforgetful nutcrackers, insects have a complex central nervous system, but it is anatomically different from that of a nutcracker or any other vertebrate. An insect's central nervous system consists of a small brain situated in the head just above the foregut, a complex of three ganglia (clusters of nerve cell bodies) in the head just below the foregut, and a ventral nerve cord that runs the length of the thorax and abdomen and may include as many as eight ganglia or, in flies, only one large compound ganglion in the thorax. In insects, control functions are not as centralized in the brain as they are in birds and mammals. A decapitated and consequently brainless insect may survive for days or weeks if its neck opening is sealed to prevent blood loss, and it may even exhibit simple behaviors. A headless cockroach can make cleaning movements. Sometimes a female mantis eats her mate during copulation, usually beginning with his head. A headless male will, however, continue to copulate and will ultimately inseminate his mate. Clearly, cleaning movements and copulation can be coordinated and controlled by some part of the nervous system outside of the head: the ganglia of the thorax and abdomen.

The behavior of insects is controlled largely by instinct, much more so than is the behavior of birds. This does not mean that insects cannot learn, that they cannot modify their behavior according to what they experience. Ants can learn to run a maze, hunting wasps memorize landmarks in order to find the way back to their nests, and honey bees can remember the location of a good foraging site for weeks. Sometimes learning by insects has a peculiarly rigid aspect. A female digger wasp may care for a dozen or more offspring at the same time, each one in a separate nest in the soil and each of a different size and thus requiring a different quantity of food. In the morning, the wasp inspects each nest, and later in the day she returns to each nest with the quantity of food indicated by her earlier inspection. However, experimental manipulation of the nests has shown that she is capable of judging the condition of a nest only during her morning visit. If, after the wasp's morning visit to a nest that is short of food, the experimenter packs into this nest more than enough food for the resident larva,

the wasp will still stuff in more caterpillars when she returns in the afternoon. She ignores what she sees on her second visit.

An insect's reproductive system generally consists of paired gonads, testes in the male and ovaries in the female. Accessory glands are usually present. In females they provide glue for sticking eggs to a surface or the material for forming an egg case, and in males they are usually involved with maintenance of the sperm and the formation of gelatinous capsules that enclose the sperm. Most females also have a pouch for storing and maintaining sperm after copulation.

Male birds, except for most waterfowl, turkeys, chickens, ostriches, and a few others, lack an intromittent organ, or penis. When a penis is lacking, fertilization is accomplished by a "cloacal kiss." The cloaca, which opens to the exterior through an anus-like vent, is the chamber into which the fecal, the urinary, and the reproductive ducts open. The sperm are transferred when the male mounts the female and the two bring their cloacas together. Then the sperm swim up the female's oviduct, where they will encounter ripe ova.

Unlike the majority of birds, almost all male insects have an intromittent organ that can be called a penis by analogy with mammals and that penetrates the "vagina" of the female to deposit sperm that are then held in the female's sperm pouch. Some male damselflies have a highly specialized penis with which they can enter the female's sperm pouch and scoop out and discard the sperm of a previous male before they deposit their own. (To achieve the same end, male dunnocks, also known as hedge sparrows, peck at the female's cloaca before mating to stimulate her to eject sperm from previous matings.) A few very primitive insects, such as the silverfish and its relatives, practice indirect fertilization. The male deposits a packet of sperm on the ground, and it is up to the female to locate it and stuff it into her "vagina."

Most birds copulate frequently before and during the period of egg laying. Perhaps the world record holder is the female Smith's longspur. As James V. Briskie, then of Queen's University in Kingston, Ontario, reported in 1992, a female copulates anywhere from 241 to 629 times for each clutch of eggs that she lays. Copulations begin from three to five days before the first egg is laid and cease a few days later, just before or shortly after the next to last egg is laid. Although other birds generally copulate often during the period of egg laying, I doubt that many of them approach the Smith's longspur in the strength of their libidos.

I do not know of any insect that copulates as frequently as do Smith's longspurs. In fact, many female insects refuse the attentions of males after they have copulated only once. Male insects, however, generally copulate often and with as many different females as possible. Some insects remain coupled for a very long time. Male and female cecropia moths normally stay coupled for about 15 hours, from just before dawn until dark, but a walkingstick holds the endurance record among insects. The tiny male, only about one-quarter the length of his huge mate, may remain coupled with her for as long as 79 days as he perches on top of her abdomen. The male thus prevents his mate from being inseminated by other males.

Some female birds store sperm for anywhere from a few days up to a record of 72 days for turkeys. A few even have special organs for the storage of sperm. But almost all female insects, with the exception of the mayflies, have a sperm pouch, or spermatheca, in which they can store and keep sperm alive for anywhere from a few days to several years. A queen honey bee can store sperm for as long as eight years. Her only opportunities to copulate come on mating flights that she makes shortly after she metamorphoses from the larval to the adult form, and during which she will mate with several males. She returns to the colony with about 6 million sperm in her spermatheca. During her tenure in the colony, she will ration out those 6 million sperm to fertilize as many as 2 million eggs. Contrast this with human females, who receive about 70 million sperm at each copulation, only one of which will—only sometimes—succeed in fertilizing an egg.

Unlike humans, all birds and most insects are oviparous—they lay eggs. But some species of insects do give birth to living young. Female aphids may give live birth to as many as seven tiny nymphal offspring each day. The female of the infamous tsetse, an African fly that transmits the protozoan that causes sleeping sickness in humans, retains her offspring within her "uterus," an expanded area of the egg passage. There the larva drinks from a "milk gland" that empties into the uterus. It is not born until it is fully grown and ready to metamorphose to the adult stage.

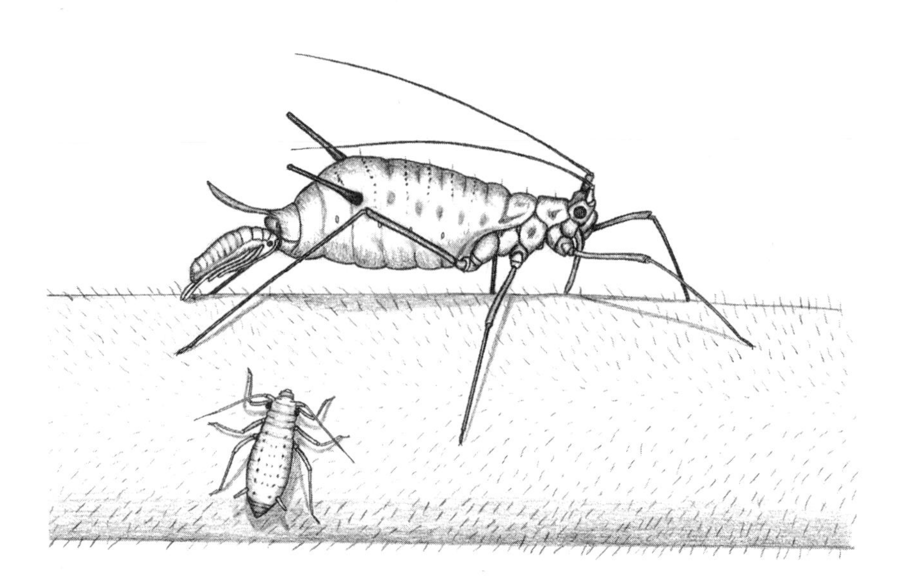

A female aphid giving birth to a nymph as an older nymph looks on

The external structure of insects is different from that of vertebrates. While birds, people, and other vertebrates have an endoskeleton, a bony internal skeleton, insects and their relatives, such as crabs, spiders, mites, centipedes, and millipedes, have an exoskeleton, an external skeleton that consists of a hard cuticle that is secreted by the layer of epidermal (skin) cells that covers the body.

The cuticle is nonliving and cannot grow. Therefore, it must be molted and replaced from time to time as the insect increases in size. It consists of three major layers: two relatively thick inner ones that are mainly skeletal in function and a very thin outer layer. Over most of the body the cuticle is hard—not at all stretchy like rubber, but elastic and flexible in the sense of a steel saw blade that will spring back to its original shape if it is bent. But over other parts of the body, especially in areas where free movement is required, as in the joints of the legs, at the bases of the wings, or between the segments of the abdomen, the cuticle is soft and membranous.

The thin outer layer actually consists of several microscopically thin sublayers, the outermost of which is a hard cement layer that protects the soft layer of wax that lies directly beneath it. The all-important wax layer prevents the loss of water vapor from the insect's body. Many insects will die of dehydration if even a relatively small area of the wax layer is abraded away. The late Sir Vincent Wigglesworth, in his day the dean of insect physiologists, designed a simple but elegant experiment that neatly demonstrated the water-retaining function of the wax layer of *Rhodnius,* a large bloodsucking true bug. *Rhodnius* engorges so fully that its heavy, swollen abdomen drags on the ground as it walks. After plugging the anuses of several of these bugs to prevent emptying of the gut, Wigglesworth allowed them to walk on a piece of filter paper dusted with alumina, a powerful abrasive. After 24 hours, the bugs died, having lost over 46 percent of their weight, presumably as water that passed through the portion of the cuticle from which the cement and wax layers had been abraded. Control bugs that walked on clean filter paper lost only 2.2 percent of their weight. To prove that the alumina acts by abrading the cuticle rather than in some other way, Wigglesworth did a second control experiment. Bugs whose abdomens could not drag because they were raised above the surface by a peg of paraffin suffered no abrasion when they walked on alumina-dusted filter paper, lost little weight, and survived for several days.

How can an insect's sense organs perceive external stimuli if the cuticle is mostly hard and altogether nonliving? The answer is threefold: the lenses of the eyes are made up of modified, transparent areas of cuticle; sounds are perceived by nerve cells that sense the vibrations that sound waves set up in thin, tautly stretched, membranous tympanic areas of the cuticle; and tiny nerve cells for the perception of touch, smell, and taste are borne in thin-walled hairs or domes that penetrate through the nonliving cuticle.

An insect's body consists of a series of segments, as does the body of any other arthropod—a segmented creature that has an exoskeleton and paired, jointed legs—or of the annelid worms that we use as fish bait and that are distant relatives of the arthropods. Most entomologists believe that the insects and the other arthropods are descended from a primitive ancestor whose body consisted of a series of similar segments, each bearing a pair of segmented legs. Through evolution, the segments of the insect body became more or less united to form three regions: head, thorax, and abdomen. The lines of segmentation are still readily visible in the thorax and the

abdomen, but the segments that form the head have become so completely fused that its segmental origin is apparent only in the developing embryo.

Three of the six head segments retained their legs, but these legs came to be grouped around the mouth opening and were modified as the mouthparts that insects use for eating as they snip, grind, or suck. The skull-like head capsule of an insect, in addition to bearing the organs of ingestion, the mouthparts, is the sensory center of the body, bearing the eyes, the antennae, and the antenna-like palpi that arise from the mouthparts. Although the lines of segmentation are still visible on the thorax, the three thoracic segments are, in most insects, solidly fused to each other so as to form a firm and immovable base for the three pairs of legs and the two pairs of wings. The abdomen consists of ten or eleven segments. In most insects, the segments of the abdomen have lost their legs, but a few very primitive, wingless insects retain recognizable vestiges of some of the abdominal legs. The legs near the rear end of the abdomen of grasshoppers, leafhoppers, bees, and wasps have been greatly modified to form genital claspers in males or an egg-laying device, or ovipositor, in females.

"A bird's bill and tongue are its key adaptations for feeding. The size, shape, and strength of the bill prescribe the potential diet." So wrote the eminent ornithologist Frank B. Gill. And, as we will see later, the same can be said of the mouthparts of insects. The bills of birds take a great variety of forms. Descriptions of only a few of them suffice to illustrate the breadth of the adaptive radiation that the bill of the ancestral bird underwent as birds came to exploit new and often more specialized diets. Carnivorous birds such as hawks and owls have powerful hooked beaks for tearing flesh and sinew. The bill of the woodcock is long, can probe deep into the soil, has organs of smell at its tip, and can spread at its tip to capture earthworms. Flamingos have oddly shaped bills that filter unicellular algae and other tiny organisms from the water. Cardinals and other finches have stout bills for cracking seeds. The long bills of hummingbirds are suited for probing deep into blossoms in search of nectar.

The mouthparts of insects are at least as varied as the bills of birds. Many insects—grasshoppers, beetles, and caterpillars among them—eat solid foods and have variously modified chewing mouthparts that are probably more or less similar to the ancestral mouthparts from which all of the other types have been derived. Two of the mouthparts, neither one of them

An adult female grasshopper

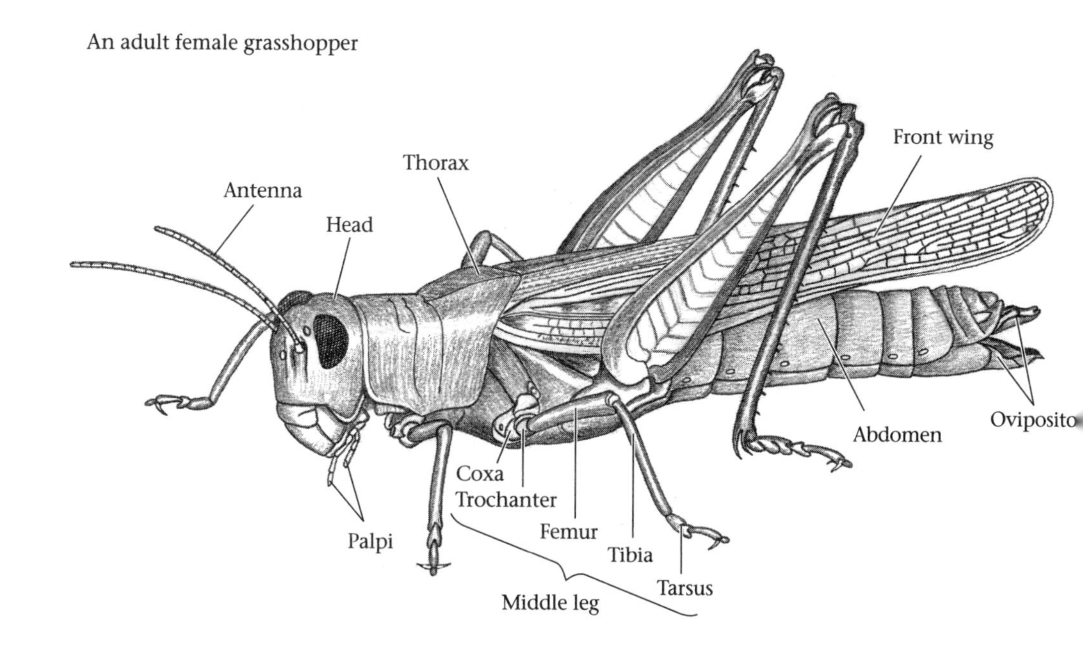


paired, are not derived from legs: the labrum, which is a flap of the head capsule ("skull") and extends down over the other mouthparts and acts as a front lip; and the hypopharynx, which lies between the other mouthparts and functions much as does our tongue. Those mouthparts that are actually modified legs are a pair of mandibles, a pair of maxillae, and the labium. The labium is actually a pair of maxilla-like appendages (sometimes called the second jaws) that have fused to form a single liplike structure. (See the drawing of the head of a grasshopper on p. 28 and the drawing of the head of a mosquito on p. 30.) In chewing insects, the mandibles are short, stout, and modified for snipping and grinding; the maxillae have a pair of antenna-like palpi, a structure for snipping, and a cuplike structure that shields the other mouthparts from the sides; and the labium serves as the back lip.

Insects with chewing mouthparts eat solid food of all sorts—of both plant and animal origin. Tens of thousands of insects, including grasshoppers, katydids, beetles, and caterpillars, use chewing mouthparts to eat plants. Different insects attack different plant organs and tissues: roots, stems, flowers, pollen, fruits, and very often leaves. And virtually all species of plants are

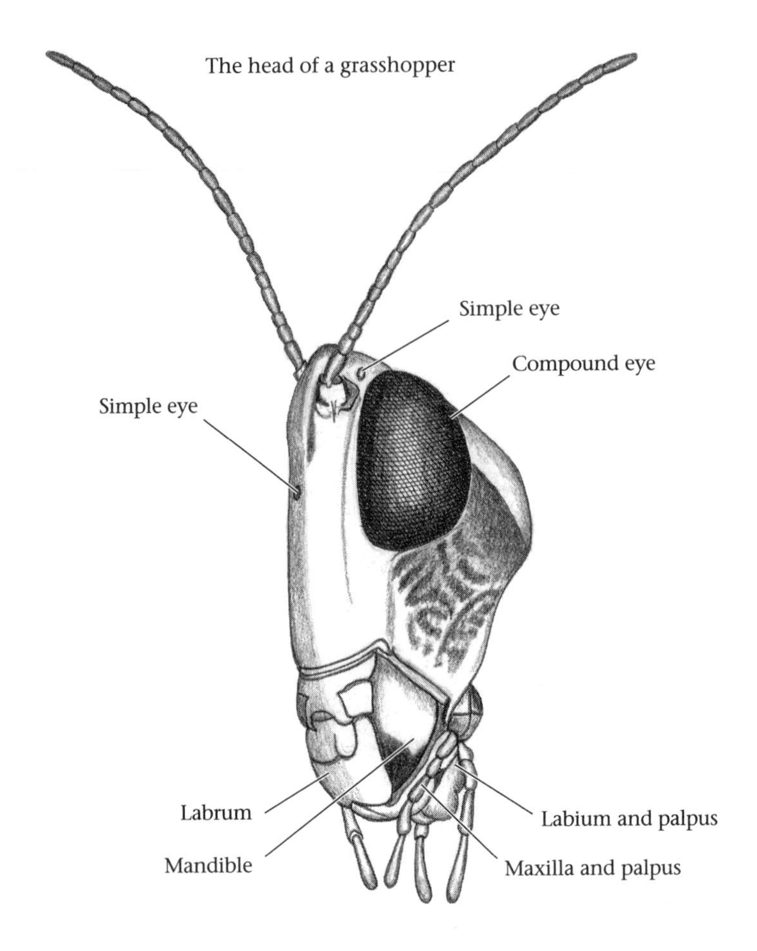


attacked by one insect or another. Some insects are specialists that will eat only certain plants—often only a few species that all belong to the same plant family. Others are generalists that attack many different kinds of plants that belong to a variety of families, but no insect is known to feed on all available plants.

Many birds exploit plants for food. The great majority of these plant feeders eat either seeds or fleshy fruits, but there are exceptions. Gallinaceous (chickenlike) birds, such as the Montezuma quail of our Southwest, use their strong feet to unearth tubers. Geese crop grass, and the hoatzin of South America and the kakapo of New Zealand, a flightless, nocturnal, and nearly extinct parrot, eat the leaves of broad-leaved plants.

Evolution designed sweet, fleshy fruits to be eaten by animals, especially birds. The plant is served if seeds are swallowed and subsequently distributed in the animal's droppings. Small, colorful fruits such as those of wild cherries, mountain ash, or pokeberry are an enticement to birds, which swallow them whole and are particularly good at seed distribution, because they may fly long distances before they eliminate the seeds in their droppings or by regurgitation. Insects such as fruit flies, blueberry maggots, and cherry fruitworms (tiny caterpillars) also exploit fleshy fruits, but they only burrow in the flesh as they feed and do not distribute the seeds.

My life-long interest in natural history was cemented by the flush of pleasure that I felt when, as a young teenager, I figured out for myself why so many of the country roadsides in Connecticut were lined by almost unbroken rows of wild black cherries. I put two and two together when I saw robins and other birds engorge themselves on ripe cherries, and then saw them defecate as they perched on roadside fences or telephone wires. Many years later, my supposition was confirmed when I read Mary F. Willson's fascinating book on the reproductive strategies of plants.

Plants have also enticed insects, especially ants, to distribute their seeds. Ants are too small to swallow fruits whole, but some plants tempt them with an appendage, the elaiosome, that is attached to the seed and is rich in fats and proteins, nutrients that ants hunger for. The ants carry the seeds back to their nests—where they may be stored for some time. They ultimately eat the elaiosomes, and then discard the unharmed reproductive part of the seed in an above-ground trash pile, where it can germinate.

Other insects take liquid foods and have their mouthparts accordingly modified. In the true bugs, such as the squash bug, chinch bug, and their close allies, and in the aphids, leafhoppers, and related insects, the mandibles and the maxillae are greatly elongated to form a tube for piercing and sucking that, when not in use for feeding, is protected by the elongated labium, which is lengthened and folded to form a trough that serves as a scabbard for the sucking tube. Different species of true bugs eat different foods, and with only minor modifications, they use mouthparts of this basic plan to suck body fluids from insects, sap from plants, or blood from humans, birds, or other vertebrates.

Like some of the true bugs, mosquitoes also suck blood from animals, especially mammals and birds, but their mouthparts, although well suited

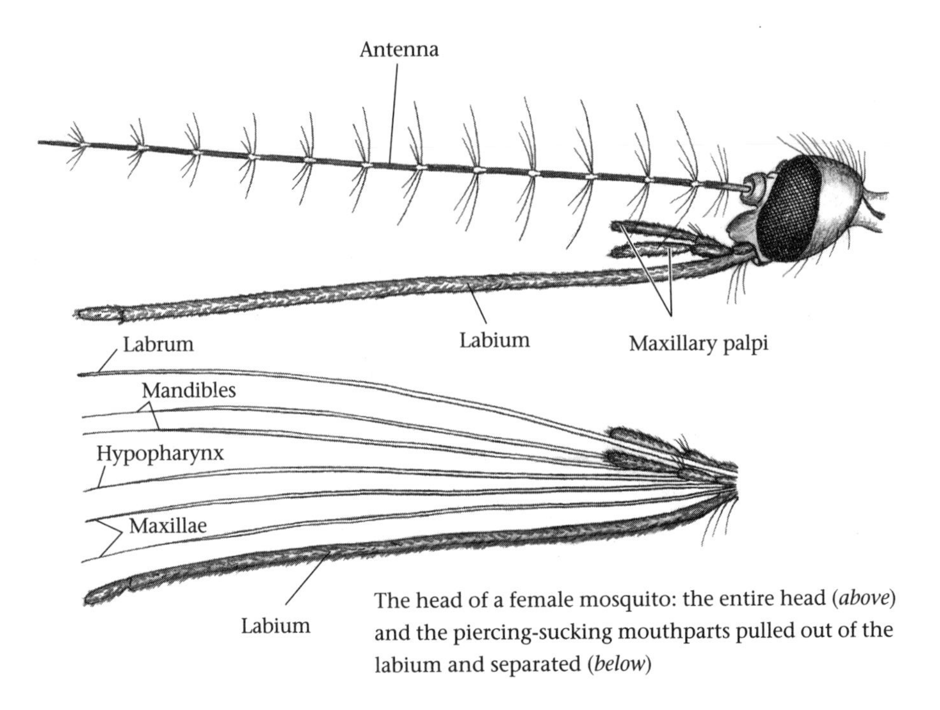


The head of a female mosquito: the entire head (*above*) and the piercing-sucking mouthparts pulled out of the labium and separated (*below*)

for piercing and sucking, are anatomically different from those of the true bugs. In female mosquitoes, the blood-sucking sex, labrum, mandibles, maxillae, and hypopharynx are all retained, lengthened, and fitted together to form a long piercing-sucking tube that is held within the elongated and scabbard-like labium when not in use.

The blossoms of animal-pollinated plants have evolved in color, shape, and size to attract and accommodate those specialized nectar feeders that serve them well by reliably distributing pollen among their blossoms and to exclude those that steal nectar without carrying away pollen and those that are not reliable inseminators because they are fickle feeders that wander from one species of flower to another as they collect nectar. Pollinating creatures, mainly insects and birds, but also bats and a few other animals, have coevolved in anatomy and behavior with the blossoms they favor. The bills of many hummingbirds have become adapted to blossoms of many different shapes and sizes. Some of these birds have very long straight bills to penetrate deep flowers. Others have shorter straight bills, and yet others

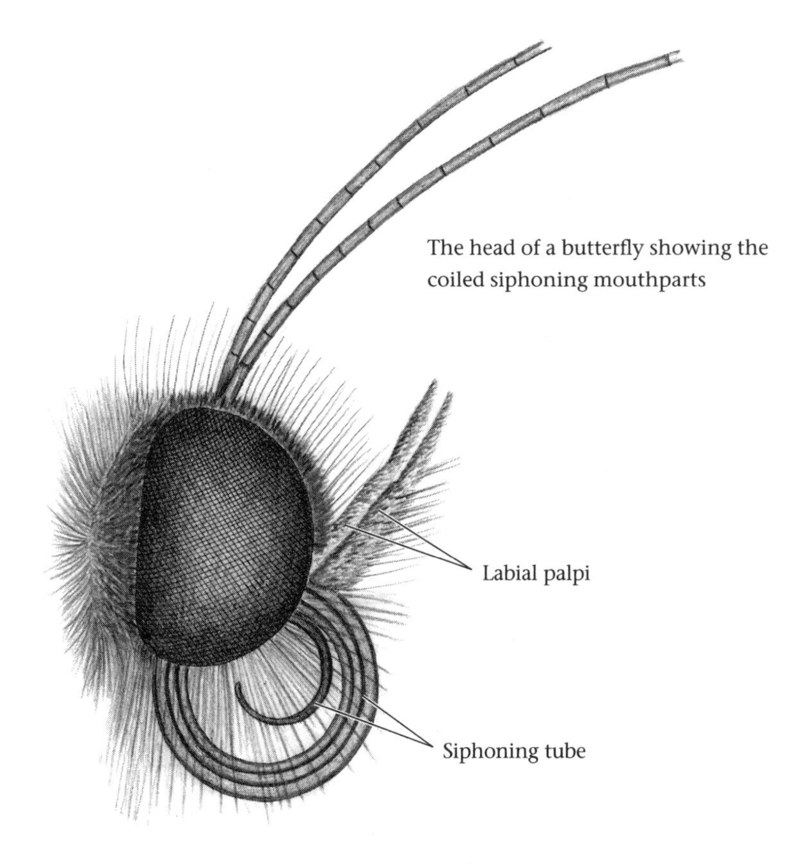


The head of a butterfly showing the coiled siphoning mouthparts

have bills that are gently or even sharply decurved to exploit variously shaped flowers.

Most of the thousands of species of butterflies and moths, important pollinators of some plants, drink nectar from flowers and have feeding organs of yet another type. Most of their mouthparts have been lost or greatly reduced, but parts of the maxillae are elongated to form a long, thin, flexible sucking tube that can, like a soda straw, reach deep into a flower to draw up nectar. No piercing is involved. When not in use, the tube, or "tongue," is coiled out of the way beneath the head and between the labial palpi. The flexible "tongue" of a moth or butterfly can enter into and sip nectar from flowers of almost any shape. But its length varies from species to species, presumably as an adaptation to flowers of various lengths. Char-

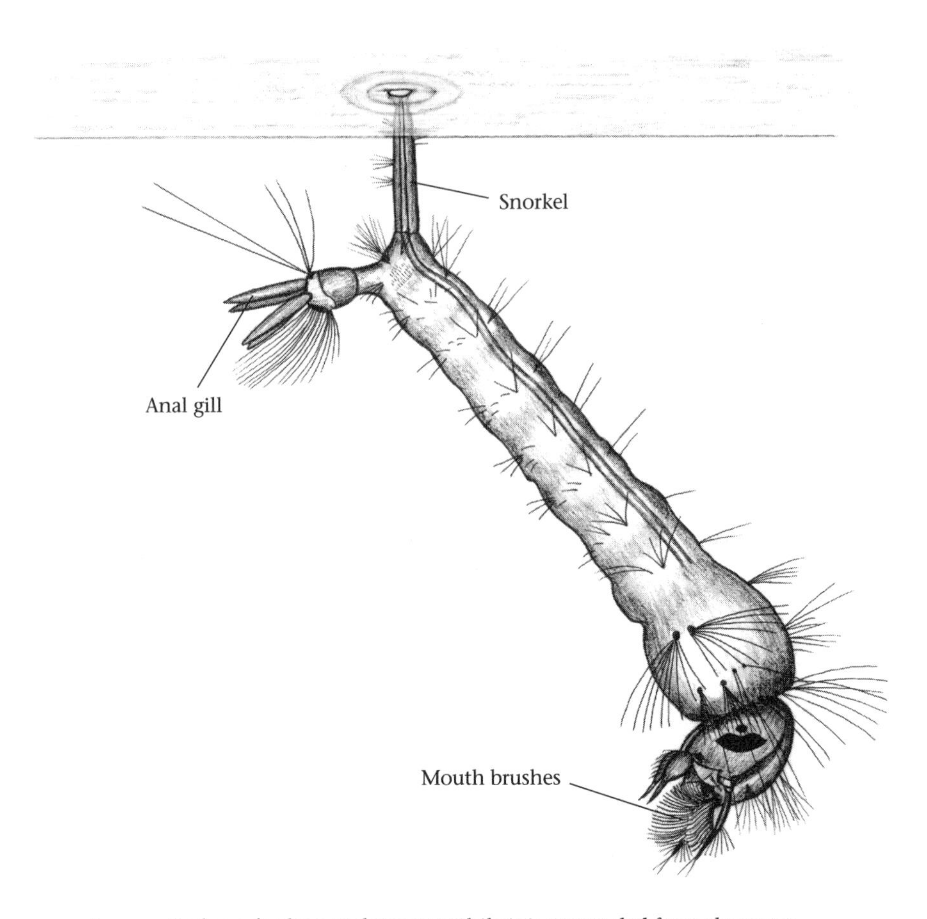


A mosquito larva feeding under water while it is suspended from the water surface by its snorkel

les Darwin knew that plants and pollinating insects coevolved to accommodate each other. Based on his knowledge of an orchid from Madagascar that has an exceptionally long blossom, he predicted that a moth with a 10-inch-long "tongue," long enough to reach the nectar at the base of this blossom, would eventually be found and shown to be the pollinator of this orchid. Since then, a moth with a 10-inch tongue has been found on Madagascar, but it has not yet been caught in the act of taking nectar from the orchid in question.

Some flies, such as blow flies and house flies, have sponging mouthparts that enable them to feed on thin films of liquid which cannot be exploited by most other insects. They have a stubby proboscis, composed partly of the labium, that ends in a pair of broad, spongelike pads. When the fly feeds on a thin film of liquid such as honeydew, the watery and sugary excrement of aphids, the pads sponge the liquid into a pool, which is then sucked up through a tube, composed of the elongated labrum and hypopharynx, that extends from the mouth down the length of the proboscis to the area between the spongelike pads. These flies can also exploit dry films of honeydew or even sugar in a bowl. They simply press the spongy pads to the dry substance, regurgitate and salivate onto it, and then sponge up this liquid along with the sugar that has dissolved in it.

One spring off the coast of Newfoundland, I watched humpback whales swimming through the sea. Although I could not see inside their jaws, I knew that the fringed baleen plates in their cavernous mouths were straining small animals from the water. As I watched them, I was struck by the realization that some birds and even some tiny insects obtain their food in essentially the same way as do the gigantic baleen whales. Some birds, notably flamingos and dabbling ducks such as northern shovelers, and some insects, such as the aquatic larvae of mosquitoes and black flies, are also filter feeders. All of these animals, from the great whales to the tiny mosquito larvae, the latter hundreds of times smaller than the shrimplike krill that whales strain from the sea, have, independently of each other, evolved different ways of straining food from water. As the mosquito larva hangs suspended from the surface film of the water by its snorkel (air tube), the brushes on its maxillae sweep like fans to form currents that are sucked into the throat, where comblike structures filter out unicellular algae and other small organic particles. The baleen plates of the whales, the fringed tongue of the shoveler, and the mouth brushes and strainers of the mosquito larva accomplish essentially the same end.

An insect's head bears not only the mouthparts, the organs of ingestion, but also the major sense organs: the eyes, the antennae, and the antenna-like appendages of the mouthparts called the palpi. These sense organs provide an insect with the senses of sight, touch, smell, taste, and sometimes even the sense of hearing. But few insects have acute vision and

some have no eyes at all. Many insects lack ears, although some earless species, such as honey bees, are acutely sensitive to vibrations of the substrate on which they sit or stand. Almost all insects depend largely upon their chemical senses, just as we rely mainly on vision and hearing. This is one of the main reasons why humans find it difficult to interpret and understand the behavior of insects.

Although some insects are eyeless, most of them have a pair of compound eyes (see the drawing of a grasshopper's head on p. 28). If you examine a compound eye with a hand lens, you can see that it is an aggregate of many transparent hexagonal lenses. The light that passes through each lens stimulates a light-sensitive element beneath it. Thus, unlike the eye of a bird or a human, in which a single lens focuses the image onto the retina, the eye of an insect is, true to its name, a compound of many separate visual units, each a light-sensitive structure provided with its own lens. Compound eyes are less adept than our own eyes or birds' eyes at resolving form and detail, but they are far quicker to perceive movement. (Just think how difficult it is to snatch a fly with your hand.) The ability of compound eyes to resolve detail does, however, increase with the number of visual units that they contain. The wingless and largely inactive female of a firefly has only 600 visual units in her compound eyes, but the winged male, which must find the female by visual means, has 5,000 in his eyes. Dragonflies, which require great visual acuity because they catch their often tiny insect prey on the wing, may have a total of 56,000.

Insects, like birds and humans, see colors—not surprising since so many of them are attracted to flowers with bright colors, especially yellow and blue. Although butterflies, like hummingbirds, are attracted to red flowers, most other insects do not see as far into the red range as do we and most birds, but unlike us and like some birds, they see into the near ultraviolet range. Not surprisingly, some flowers have markings that are visible only in the ultraviolet range and that often act as "nectar guides" indicating where the nectar is.

Many insects, especially winged species, also have two or three "simple eyes" that lie between or above the compound eyes. Each simple eye, or ocellus (pl. ocelli), consists of a single large lens with a cluster of light-sensitive elements behind it. The simple eyes are not well adapted for resolving images. Although those of some insects can form an image, it is focused far behind the light-sensitive elements and is thus badly out of focus. The simple eyes, however, are more sensitive to light, especially dim light, than are the compound eyes. As Sir Vincent Wigglesworth put it in terms familiar to photographers, the compound eyes of certain ants have an aperture of only *f*/4.5, while the simple eyes are "faster" lenses with a much larger aperture of *f*/1.5. What role, then, do the simple eyes play in the life of an insect? Although we do not have all the answers to this question, it seems certain that they help regulate the activity rhythms of some insects whose behavior is at least partly governed by the daily cycle of light and dark. Bees with the simple eyes painted over and the compound eyes left uncovered are much slower to wake up in response to light than are bees with both the simple eyes and the compound eyes uncovered. American cockroaches normally maintain a daily activity rhythm, with most activity in the dark, that is governed by light intensity. If the simple eyes are blackened, this rhythm is disrupted even if the compound eyes are left uncovered.

Almost all insects, both adult and immature forms, have a pair of antennae on the head. Depending upon the species of insect, the antennae are responsible for one or more sensory functions: touch, taste, smell, sound reception, moisture sensing, and temperature assessment. The antennae are often long, threadlike, and many-segmented, but in quite a number of insects their form is greatly modified to facilitate particular sensory functions. The short, three-segmented antennae of blow flies, house flies, and related species function as wind-speed indicators. The bulbous third (last or terminal) segment bears a featherlike projection that catches the wind like a sail and thus causes the entire segment to twist. The central nervous system translates the degree of twist into wind speed or the air speed of a flying individual. In some other insects, notably saturniid moths—the giant silkworms such as cecropia, polyphemus, and promethea—the antennae of the males are greatly enlarged and featherlike in order to accommodate an enormous number of odor receptors. The males use these to perceive an airborne sex-attractant pheromone, a chemical signal, that is emitted by the females of their species. When a male smells this pheromone, he flies upwind, often for a considerable distance, until he finds the female. Male mosquitoes also use their antennae in locating females, but in their case the bushy antennae function as sound receptors that are set to vibrating by the sound waves, perceptible to us as a humming tone, that are produced by the

beating wings of a female. Unmated male mosquitoes confined in a cage without females are driven to a sexual frenzy by a tuning fork that vibrates at about the same pitch as do the wings of flying female mosquitoes.

Although pheromones with various functions are important avenues of communication for insects and mammals, birds are not known to produce sex-attractant pheromones or, for that matter, pheromones of any sort. But, as with mosquitoes and some other insects, sounds are an all-important medium of intersexual, intrasexual, and even interspecific communication for most birds. Sound is produced by the vibrations of the air column as it flows through the narrow passages of the bird's syrinx, which is located in the body where the trachea, the main air tube leading to the lungs, branches to form the two major bronchi, one leading to each lung.

With the exception of cicadas, most insects make sounds much as do the crickets that you met at the beginning of this chapter. The male cricket produces his chirp by stridulating, rubbing a scraper on one forewing against a filelike structure on the other, much like rubbing a rasp against a piece of sheet metal or running a thumbnail along the teeth of a comb. The male grasshopper produces a softer and more rasping sound by rubbing a file on the hind leg against the front wing. Like the song of a bird, the cricket's chirp or the grasshopper's rasp invites females of the species to join him in his territory and warns other males to stay away.

But not all males stay away. Near a singing male cricket may lurk sneaky "satellite males" that remain silent although they do have sound-producing organs. Sometimes these sneaky males manage to intercept and inseminate a female as she approaches the singing male. It is no different with at least some birds. An attractive singing male holding a prime territory may be cuckolded by other males who surreptitiously copulate with his mate, as is the case with the polygynous red-winged blackbird. A male red-wing sings his exuberant *konk-er-ree* as he defends his territory and a harem of as many as a dozen females against intruding males. But as was demonstrated in 1975 by the research of Olin Bray and his colleagues, he is not always successful in assuring the fidelity of his mates. In an experiment that may seem heartless to some people, they vasectomized some of the territorial male red-wings in a small marsh. The now sterile males continued to copulate with their wives, but some wives laid fertile eggs nonetheless, showing that they had sneaked copulations with the still fertile rulers of adjacent territories or with unobtrusive, nonterritorial males, the avian equivalent of satellite male crickets. Fifteen years later, in 1990, H. Lisle Gibbs and several colleagues used DNA markers to again demonstrate that red-winged blackbirds are promiscuous. They showed that about 20 percent of the chicks fathered by an average male were produced by females other than those of his own harem.

Male cicadas, including periodical species that become adults and emerge from the soil only once every 13 or 17 years, make sounds not by stridulating but by a different method. Their sound-producing organs, anatomically the most complex of the animal kingdom—more complex than the syrinx of a bird or the larynx of a human—are located in two cavities on the underside of the body near the base of the abdomen. Sound is produced by a springy, domelike organ called a timbal. A pull of the muscles attached to the timbal causes it to buckle inward and make a click—like pressing down on the top of a can. When the muscles relax, the elastic timbal makes another click as it springs back outward. A male cicada makes about 390 clicks per second to produce what to our ears sounds like a high-pitched squawk.

The song of the male periodical cicada calls in females, but, unlike the song of a bird or a cricket, it also calls in other males rather than warning them away. A single male may be the nucleus of a chorus of males that may grow to include hundreds or even thousands of individuals—all singing the same song in unison. Large choruses are more attractive to females than are individual singing males, probably because they are louder. A male is more likely to find a mate if he joins a chorus rather than going it alone.

Periodical circadas occur by the millions and make a deafening uproar. In the spring of 1990 I visited my daughter and her husband in the northern suburbs of Chicago during the height of the emergence of brood XIII, one of 15 "broods" of periodical cicadas that occur in the United States. (A brood is defined as all of the 13- *or* 17-year cicadas that emerge in the same year.) The cicadas were so noisy that we had to raise our voices when we were talking outdoors. Their din drowned out the songs of even the noisiest birds. I watched a robin but heard nothing although its open beak and throbbing throat showed that it was singing.

Each of the three segments of the thorax of an insect bears a pair of legs, except in some immature insects that are completely legless, such as fly maggots and the larvae of fleas, bees, and wasps. Like the legs of birds, the legs of insects have been variously modified to perform a variety of functions. Although those of most insects are designed for walking or running, as in most cockroaches and adult beetles, flies, and butterflies, those of other insects are modified for other types of locomotion or sometimes, as in praying mantises, even for functions that have nothing to do with locomotion. The legs of some insects, such as those of whirligig beetles and the true bugs known as water boatmen, are broadened like paddles and used for swimming. The hind legs of grasshoppers, crickets, and katydids are obviously modified for jumping, while the first and middle pair are of the walking type. The coxa, the trochanter, and the tarsus of the jumping hind leg are not specially modified (see the drawing of the grasshopper on p. 27), but the femur is swollen to accommodate the large mass of muscle that causes the elongated tibia to thrust downward as the insect leaps away to avoid an enemy.

The legs of water striders are suited for skating on the surface film of a quiet body of water as they glide about in search of their prey, small insects that fall onto the surface of the water. The legs do not penetrate the surface film because the sharp claws near their tips are retracted in depressions and because the tarsi are clothed with oily, water-repelling hairs. In his beautifully illustrated *Insects of the World,* Walter Linsenmaier described the shadows cast on the bottom by the dimples made in the surface film by the legs of a water strider as it skated over the water's surface:

> When sunlight bewitches the brook, casting the reflection of sky, thickets, and trees in cool colors onto the surface and making the water below luminous with its warm light, here and there it traces on the bottom silhouettes [shadows] composed of symmetrical spots. At one instant they are still or flowing slowly along, then suddenly they dart jerkily away in ghostly, disembodied designs over the golden yellow of the bottom, the copper of sunken leaves and the shining green of the water plants.

The legs of the northern mole cricket are quite different, suited for digging. A species native to North America, this cricket is so named because it

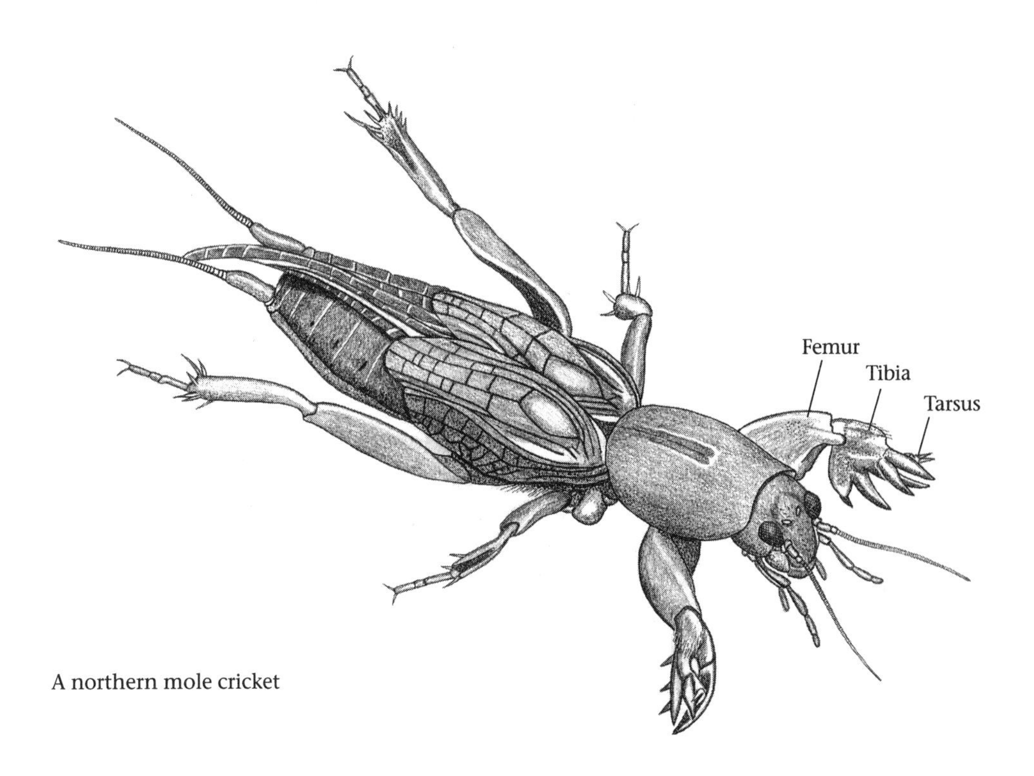


A northern mole cricket

looks and behaves very much like a miniature mole. The front legs are stocky and broadened, especially the shovel-like tibias with their rakelike spines. The tibias, driven by the powerful muscles of the femur, propel this insect through the soil as it does its subterranean breaststroke in search of worms and insects to eat.

Unlike many insects, praying mantises do not ordinarily use their front legs for locomotion. Their front legs are raptorial, even more elegantly modified for grasping prey than the talons of a hawk or an owl. Mantises do not stalk or chase their prey as do raptorial birds and some other insects. They are, instead, beautifully camouflaged ambushers that lie in wait quietly until some unwary insect comes within striking distance of their deadly front legs. The tibias and femurs of these legs bear sharp teeth, oppose each other, and, like a nutcracker, can hold the prey in a firm grip while it is being devoured. The mantis's reach is greatly extended by the elongated coxae of the front legs and by the first segment of the thorax, which is not

only elongated but, unlike the thoracic segments of most other insects, is also freely movable on the preceding thoracic segment and thus capable of swinging the whole grasping apparatus from side to side to strike at prey that would otherwise be out of reach.

The wings of birds are modified front legs. You know this very well if you have ever eaten a chicken wing. The "upper arm," the drumstick of the wing, bears no flight feathers, but to it attach the strong breast muscles that power the upstrokes and downstrokes of the wing. The secondary flight feathers, including the so-called tertials, attach to the next joint of the wing, the "forearm." The primary flight feathers are borne by the bony and much reduced "hand," the part of the wing that is discarded when Buffalo wings are prepared.

As you have already seen, the wings of insects are not modified legs but are actually flanges that protrude from the middle and last segments of the thorax. An insect wing is actually an outpouching of the body wall, a thin, flattened envelope, each of whose two walls consists of epidermis and its attendant cuticle. The wing is traversed by a network of externally visible thickenings, "veins" that distribute blood and air (the latter carried by tracheas) to the living epidermis, and that contain nerve fibers that run to the sense cells of the wing. The wings of insects are served by two types of muscles: direct wing muscles that attach to the wing at its base and indirect wing muscles that do not attach to the wing at all. In all insects, direct muscles are involved in folding the wings and controlling the angle at which they "attack" the air in flight, but in a few insects, such as dragonflies and cockroaches, they are also involved in powering the downstroke in flight. The indirect muscles, responsible for the beating of the wings in most insects, attach to the springy walls of the thoracic segments that bear the wings. These powerful muscles cause the wings to beat by moving the thoracic walls to which the wings attach.

Most insects have two pairs of wings, usually membranous, that may be variously modified. In beetles, the forewings are hard and horny, serving as covers for the hind wings and as dorsal body armor rather than for flight, as do the membranous hind wings. The wings of moths and butterflies are actually membranous and transparent like the wings of most insects, but

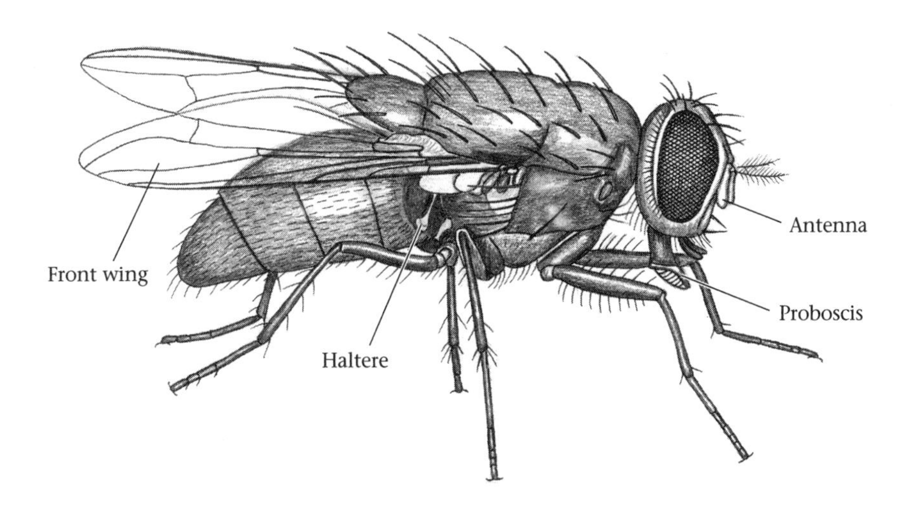


Side view of a blow fly showing the front wings and one of the halteres

they are made opaque by the minute scales that cover them, overlapping like the shingles on a roof. The colors of the wing reside in the scales, as can be seen by the colored "powder" that rubs off on our fingers when we handle a moth or a butterfly. Flies, mosquitoes, gnats, and their relatives have a pair of membranous front wings, but the hind wings have been reduced to minute, club-shaped organs called halteres. The halteres beat in unison with the wings and act as gyroscopes that sense when the insect is not in the normal flight path—when it pitches, yaws, or rolls.

The abdominal segments of most insects are movable upon each other, each ringlike segment joined to each adjacent segment by a ring of membrane that is very wide in some species. The hard upper and lower walls of a segment are joined by lateral membranous walls that are often very broad. These intersegmental and intrasegmental membranes, which are usually folded away out of sight, permit considerable flexibility and movement. The ringlike intersegmental membranes allow the abdominal segments to telescope into each other and to be extended according to need, as when a female grasshopper must lengthen her abdomen as she digs down with the ovipositor at the end of her abdomen to place her eggs deep

in the soil. The intrasegmental, membranous, lateral walls of a segment allow it to expand and contract. Together the intersegmental and intrasegmental membranes allow the abdomen to balloon out as a gravid female grows heavy with eggs or as a blood-sucking individual engorges itself.

Ovipositors, modifications of certain legs of the ancestral arthropod, may be specialized in various ways. The female grasshopper's ovipositor is short and broad for digging in the soil. Sawflies have sawlike ovipositors that cut slits into plant tissues. Most parasitic wasps have needlelike ovipositors suited for injecting an egg into the body of the host insect in which their larvae will live and feed. Some parasitic wasps have thin, flexible ovipositors that are longer than the body and are used to drill through wood to a tunnel made by a host insect. In bees, predaceous wasps, and some ants, the ovipositor has been converted to a venom-injecting stinger, used mainly to paralyze prey by most wasps, but deployed in defense of the nest or colony by bumble bees, honey bees, ants, and social wasps.

As I have already mentioned, an insect must molt its cuticle—shed its "skin"—if it is to grow to full size. Although membranous areas of the cuticle unfold and thus accommodate a modicum of growth, hard areas—notably the "skull," or the head capsule, and the legs—ultimately limit increases in size. And insects do undergo tremendous increases in size. The big green hornworm caterpillars that eat tomato leaves molt four times as they increase their weight by a factor of over 6,000 from the time when they first hatch out of the egg until, less than a month later, they attain their maximum weight as full-grown caterpillars ready to molt to the pupal, or transformation, stage.

Almost all insects undergo some degree of metamorphosis as they grow. Just as the tadpole becomes a frog, the grub becomes a beetle, the maggot becomes a fly, and the caterpillar becomes a moth or a butterfly. But not all insect memamorphoses are as dramatic as these. The most primitive insects, such as springtails and silverfish, do not metamorphose as they grow. Dragonflies, grasshoppers, lice, and true bugs, among other insects, undergo gradual metamorphosis, passing through three life stages: the egg; the nymph, or growing stage; and the adult, or reproductive stage. Nymphs and adults look and behave very much alike. If you see a tiny wingless grasshopper that has just hatched from an egg you would probably recog-

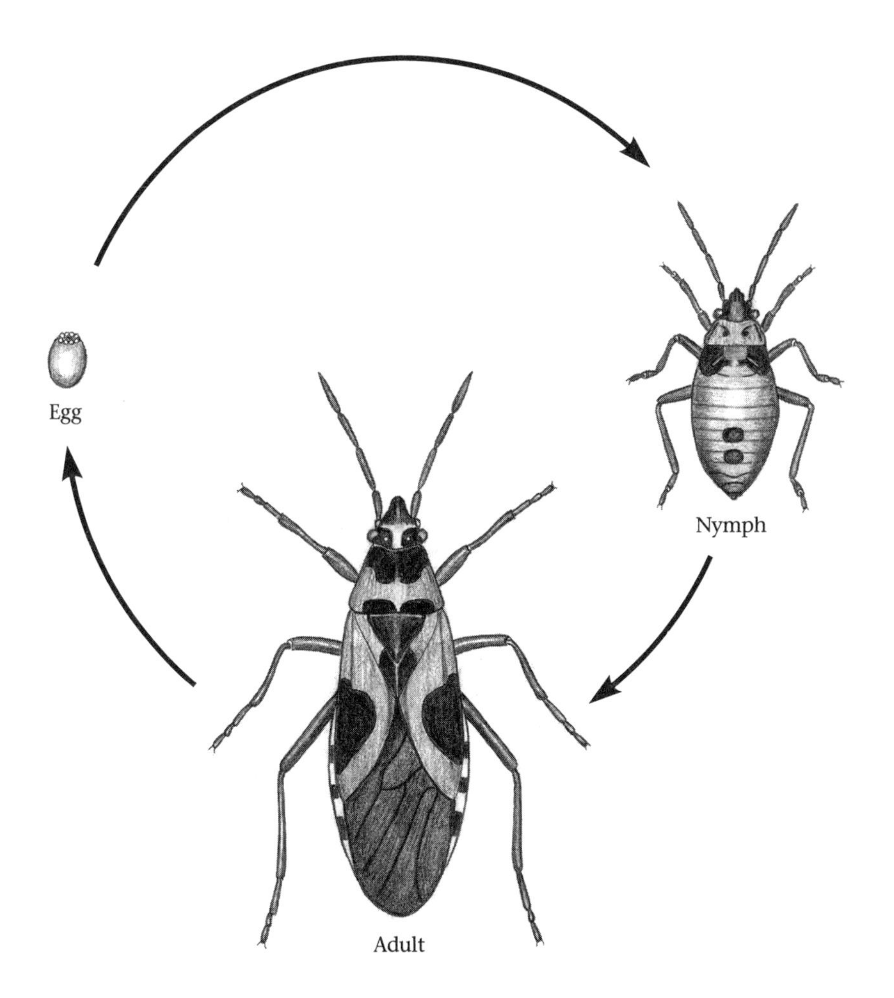


The life stages of the milkweed bug, an insect with gradual metamorphosis

nize it as the immature form of its parent. Like its parent, it lives in and among plants, it is an active jumper, and it uses its mouthparts to chew leaves. As it goes from nymphal molt to nymphal molt, external wing buds appear and grow larger with succeeding molts until, after the last one, the molt to the adult stage, they are fully developed and finally capable of sustaining the grasshopper in flight. Except for its size and fully developed wings and genitalia, the adult does not differ significantly from the nymph in appearance or behavior.

Most insects, among them flies, butterflies, beetles, fleas, wasps, ants,

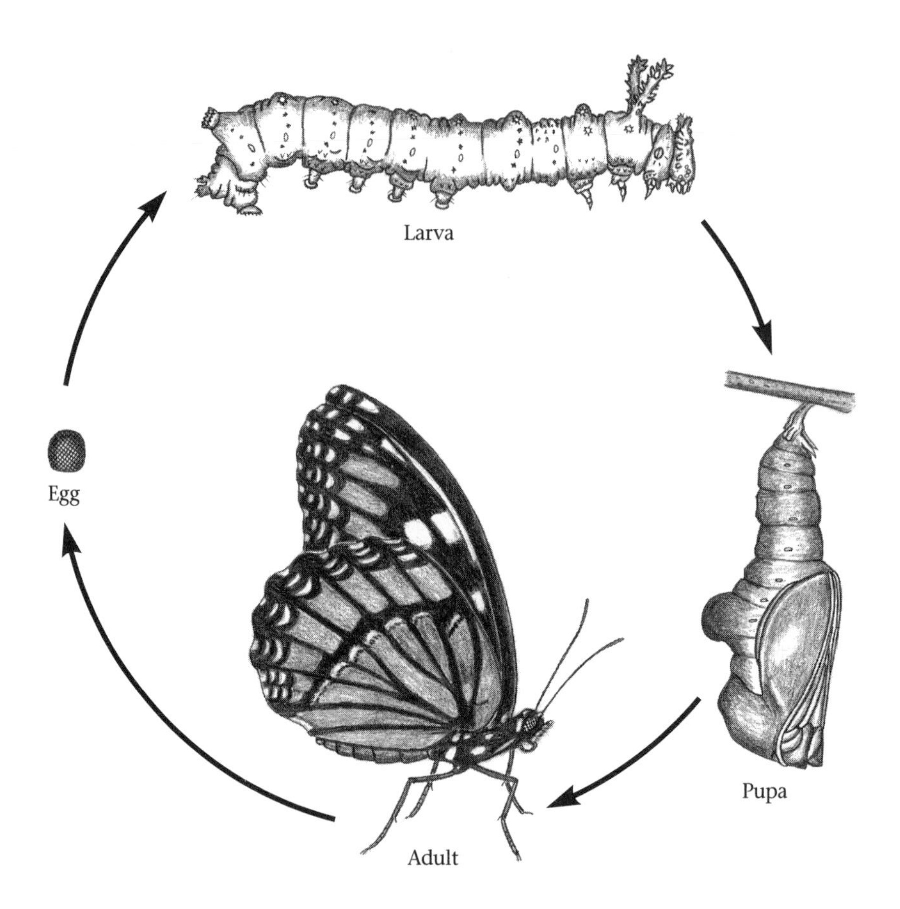


The life stages of the viceroy butterfly, an insect with complete metamorphosis

and bees, undergo a complete metamorphosis that includes four life stages rather than only three: the egg; the larva, or growing stage; the pupa, or transformation stage; and the adult, or reproductive stage. The wingless larva often bears little or no resemblance to the adult in diet, behavior, or appearance, and undergoes only slight externally visible changes as it passes through its several molts, culminating in the molt to the pupal stage. The radical transformation that occurs in the pupal stage makes it possible for larvae and adults to follow two different evolutionary paths. The soil-dwelling, root-feeding bean leaf beetle larva, or grub, becomes an aerial adult that eats the leaves of bean plants. The horn fly lives in and feeds on cow pats as a larva, or maggot, but sucks the blood of cows after it molts to the adult stage.

But the most spectacular transformation is from caterpillar to moth or butterfly. The raison d'être of every caterpillar, or larva, is to eat, grow, and survive. It is essentially a digestive system on a caterpillar tread, using its six short thoracic legs and its ten stubby abdominal prolegs to hold on tightly to its food. By contrast, an adult moth or butterfly has enormous wings covered with colorful scales, only six long, thin legs that serve mainly as the landing gear, and, instead of chewing mouthparts, a long, coiled sucking tube that can be extended deep into a flower to reach the energy-rich nectar that fuels the muscles that drive the wings. As the caterpillar is specialized to eat and grow, the butterfly is specialized to seek a mate or to fly from host plant to host plant as it distributes its eggs. Indeed, butterflies have been called "winged gonads." In insects with gradual metamorphosis we see a minimum of such specialization. Except for the fully developed wings and genitalia of the adults, the body form and life style shared by nymphs and adults are a compromise that must serve both for growth and reproduction.

Considering the efficiencies that stem from specialization, it should come as no surprise that insects with complete metamorphosis, and, hence, with a larval stage specialized for feeding and an adult stage specialized for reproduction, have been far more successful—at least as judged by the numbers of species that have evolved—than those which have gradual metamorphosis, and thus lack specialized life stages. Although there are well over 755,000 known species of insects with complete metamorphosis, there are only about 150,000 with gradual metamorphosis.

# Bugs That Birds Eat

# 3

Fleeing persecution in Illinois, the Mormons moved west to the valley of the Great Salt Lake in what is now Utah. A vanguard of 170 arrived in July of 1847 and soon planted grain. But the crop was destroyed by hordes of large, brown, wingless katydids—now inappropriately known as Mormon crickets—that descended to the valley from their breeding grounds in the surrounding hills. Today Mormon crickets still migrate down from the hills throughout much of the western United States, crawling overland slowly in swarms that may cover a square mile and include from 100 to 500 "crickets" per square foot. Voracious feeders, they will eat over 250 species of range plants and nearly all field and vegetable crops.

In 1848 the main band of Mormon settlers arrived, and a second crop was planted. Once again, hordes of Mormon crickets appeared and began to eat the grain. The settlers were threatened with starvation. But, as if by a miracle, large flocks of California gulls, which nest in nearby marshes, flew to the fields and saved the crop by eating the "crickets." In 1913, the Mormons commemorated their rescue from starvation by erecting a golden statue of the California gull in Temple Square in Salt Lake City. Over the years, Mormon crickets have repeatedly reinvaded the valley of the Great Salt Lake, and on several occasions California gulls have again come to the rescue.

Mormon crickets are not the only insects that occur in vast hordes, and many species of birds, probably the great majority, are opportunists that, like California gulls, take advantage of the occasional insect outbreak to gorge themselves. In 1907, Edward H. Forbush, the state ornithologist of Massachusetts, described how birds responded to the great swarms of grasshoppers that had plagued the Mississippi Valley in the previous century: "When large swarms of locusts appeared, nearly all birds from the tiny Kinglet to the great Whooping Crane, fed on them. Fish-eating birds, like the Great Blue Heron, flesh-eating birds, like the Hawks and Owls, shore birds, Ducks, Geese, Gulls,—all joined with the smaller land birds in the general feast."

Accounts of birds alleviating the damage done to crops by outbreaks of pest insects are numerous, especially in the writings of the economic ornithologists of the late nineteenth and early twentieth centuries. In *The Practical Value of Birds,* Junius Henderson reports that in Australia in 1894 crows gathered by the thousands to feast on swarms of grasshoppers. For a month they fed only on these insects, eating them by the millions, their stomachs containing an average of about a hundred grasshoppers each. In 1891 there had been a similar gathering of crows in response to a plague of grasshoppers in New South Wales, and in 1892 flocks of starlings, spoonbills, and cranes in Victoria had saved crops from serious injury by grasshoppers.

In 1920, W. L. McAtee, a leading economic ornithologist of his day, recounted scores of instances in which birds "came to the rescue" when insect outbreaks threatened crops. He quoted from a letter sent to him by an observant correspondent from Fresno, California:

> One spring vast numbers of rose beetles *(Aramigus fulleri)* invaded the country about Clovis (California), and after destroying the rose flowers they took to the vineyards, doing considerable damage to the foliage by boring numerous holes through the leaves, causing them eventually to wither up and drop off. Every day for nearly a week a great flock of Brewer blackbirds hovered over a certain vineyard that I had an excellent opportunity to observe. Crawling over the branches or alighting on the topmost shoot, these black-plumaged birds were conspicuous objects against the green of the tender new foliage. As a result of the efforts of these birds, in a short time the vineyard was almost entirely free from the beetles.

Large numbers of birds assembling to exploit a plague of insects—gulls eating Mormon crickets, crows eating grasshoppers—are obvious to all observers and leave a lasting impression on the mind. But in the great ecological scheme of things, these are rare and isolated events. Figuratively speaking, they rate just a few paragraphs in the many volumes—most of them yet to be written—that will tell the story of birds eating insects. The greater part of that story will describe and analyze the ordinary, everyday struggle to obtain food. Finding, capturing, and eating insects is the "profession" of an

insectivorous bird, accomplished with great expertise, central and essential to its life, and usually occupying most of its day. To understand the roles that insectivorous birds play in their native ecosystems, we must know what insects they eat and how, when, and where they capture them.

Although the limpkin and the snail kite of the Everglades eat only the large, aquatic *Pomacea* snails, I know of no bird that eats only one kind of insect, although some specialized tropical species eat insects of only one or two families. Birds generally eat a wide variety of insects, as can be seen from the analyses of stomach contents summarized by Alexander Martin and his coauthors in 1951. The northern cardinal, an omnivore, is a good example. Arthur Cleveland Bent reports that the animal portion of its diet, which consists almost entirely of insects, includes 51 species of beetles, among them ground beetles, click beetles, wood borers, and fireflies; 12 kinds of Homoptera (an order that has no inclusive common name), including cicadas, leafhoppers, aphids, and scale insects; 4 species of grasshoppers and crickets; and various species of termites, caterpillars, ants, flies, sawflies, and dragonflies. Almost all kinds of insects are eaten by birds, and the great majority of birds, except marine species, include at least some insects in their diet.

Throughout their lives, doves and pigeons are generally strict vegetarians that eat seeds and fleshy fruits, although, as A. W. Schorger points out, passenger pigeons ate caterpillars during outbreaks. Otherwise, only traces of animal matter, probably inadvertently swallowed, are found in the stomachs of pigeons and doves. But many other North American birds that are generally thought of as seed or fruit eaters, such as sparrows and finches, regularly include at least a small percentage of animal matter, usually insects, in their diet. Other birds divide their dietary intake more or less evenly between insects and vegetable matter. Among them are chickadees, titmice, crows, grackles, and blackbirds. Some other groups of birds, such as cuckoos, flycatchers, swallows, vireos, warblers, orioles, and tanagers, are largely insectivorous, but may eat fruit or nectar during migration or on their tropical wintering grounds. Relatively few birds, notably nightjars and swifts, are exclusively insectivorous.

With very few exceptions, among them doves and pigeons, even the most dedicated fruit and seed eaters, such as buntings, longspurs, grosbeaks, and some sparrows, raise their nestlings on a high-protein diet consisting mostly of animal matter, mainly insects, rather than the largely vegetarian diet that they themselves eat. (The American goldfinch, exclusively a seed eater, is another exception. It feeds its young regurgitated seeds.) Studies of common grackles have documented in detail the animal-rich diet of nestlings. After examining the stomachs of 2,346 adult common grackles, F. E. L. Beal reported that during the period of a year their food intake was 69.7 percent vegetable and only 30.3 percent animal, the latter consisting largely of insects. But these grackles feed their nestlings a diet consisting mostly of animal matter. In another study, W. J. Hamilton, Jr., found that the stomachs of 130 common grackle nestlings contained only 6.4 percent vegetable matter—green grass, grain, and fruit; 4.3 percent grit; and an overwhelming 89.3 percent animal matter, consisting mostly of insects and other arthropods.

In 1950 Josselyn van Tyne watched a cardinal that clearly distinguished between the seeds that it ate and the worms (caterpillars) that it fed to its nestlings:

> The adults regularly frequented my yard, gathering much of their food there. At noon on May 24 the adult male, on his way back to the nest territory, stopped at my feeding shelf with his beak full of small green worms such as I had often seen him feed to the young. He immediately put the worms down on the shelf and began cracking and eating sunflower seeds. After a minute or two he took the worms in his beak but again laid them down and ate a few more seeds. He then picked up the worms for the second time, flew across the street, and (presumably) fed the young. At 5:30 P.M. the same day I saw the whole incident repeated without noticeable variation.

✱ Birds and insects are members of complex ecosystems, communities of interacting plants and animals plus the physical environment—rocks, soil, air, and water. A community can be diagramed as a broad-based pyramid of feeding levels, the lowest consisting of green plants, the only organisms that can capture the energy of the sun, bind it in carbohydrates, and thus make it available to animals. By eating plants and each other, the animals garner and transfer the energy captured by plants, passing it up through the

A male cardinal holding a caterpillar in its beak

succeeding feeding levels of the pyramid, from herbivores to intermediate predators and finally to its apex, the top predators—such as mountain lions, goshawks, and great horned owls. (The situation is complicated by omnivores, which eat both plants and animals.) Most of the herbivores are insects, although there are other plant-eating invertebrates, as well as plant-eating vertebrates, ranging from birds to mice and elephants. In turn, insects, birds, mammals, and other flesh-eating creatures prey on the herbivores. Thus, either directly or indirectly, all animals depend upon plants for their food.

The community of birds in a North American forest, according to John Terborgh, is likely to include about 40 breeding species. I know of no comparably reliable estimate of the species composition of a North American community of forest arthropods, but there is no doubt that the arthropods greatly outnumber the birds. There are far more species of both birds and insects in tropical forests: over 200 breeding birds were reported from an Amazonian forest by Terborgh in the 1990s, and over 40,000 species of insects were estimated to exist in a single lowland plot in a mature Panamanian forest by Terry Irwin in the 1980s.

One way to get a broad picture of avian eating habits is to group birds, which have evolved an amazing variety of anatomical and behavioral adaptations to facilitate their own particular modes of feeding, in feeding guilds. These guilds, not unlike medieval European trade guilds, consist of groups of species that use the same class of resources in a similar way. Ecologists have recognized a number of guilds of insect-eating birds, including, among many others, *leaf gleaners,* such as warblers that pick insects from foliage; *bark gleaners,* which hunt for insects on the surface of bark, as do creepers and nuthatches; *wood and bark probers,* in North America, woodpeckers that dig in wood to find hidden insect prey; *air salliers,* which take off from a perch to snatch insects from the air, as do most flycatchers and, on occasion, many other North American birds, including cedar waxwings and Townsend's solitaires; and, finally, *gleaners of aerial plankton,* such as the swallows, swifts, nightjars, and nighthawks that capture the insects and other creatures that fly or float in the air.

The concept of the guild is helpful, but it is a simplification of reality, an abstraction whose limits are arbitrarily set. Few North American insect-eating birds use only one method to catch their prey. Most are versatile and have a repertoire of several prey-seeking behaviors, and thus do not fit neatly into one guild. The American redstart is certainly a candidate for the leaf-gleaning guild, but, like many other leaf-gleaning warblers, it often darts out like a flycatcher to snatch an insect from the air. The even more versatile yellow-rumped warbler is an adept leaf gleaner but also gleans bark, feeds on the ground, and catches flying insects, and in winter, as you read above, eats the wax-covered berries of bayberry and wax myrtle. P. M. Silloway of Lewiston, Montana, wrote of the bark-gleaning red-breasted nuthatch: "June 9 [1906] . . . at times [it] acts like a real flycatcher. Just now one alighted on a tree trunk near me, and while investigating the bark crevices, twice . . . flew out from the trunk, captured a flying insect dexterously in the air, and returned to his gleaning on the bole." Among the wood and bark probers, the northern flicker commonly descends to the ground to

catch ants and is known to glean leaves; the red-headed woodpecker also gleans foliage and captures flying insects as adeptly as a flycatcher. The least flycatcher usually snatches insects from the air or the underside of foliage, but will also scramble about on the trunk of a tree as it catches insects like a creeper and will even steal insects from spider webs.

The leaf-gleaning guild includes chickadees, titmice, kinglets, gnatcatchers, vireos, wood warblers, tanagers, orioles, and other birds that move from twig to twig as they search the foliage of trees and shrubs for caterpillars, moths, sawflies, beetles, katydids, flies, aphids, spiders, and other arthropods that eat leaves or just rest on them. Each of these leaf gleaners has its own peculiarities. Chickadees often hang upside down as they examine a spray of leaves. Gnatcatchers move nervously as they forage, flitting their tails and sometimes, like a kinglet, hovering in front of a leaf to snatch some otherwise unreachable morsel. In spring, particularly during the first two weeks of May in central Illinois, I frequently see mixed flocks of migrant warblers—often a dozen or more species, including blackburnians, Cape Mays, blackpolls, and bay-breasteds—move through the trees, usually large oaks, as they pluck insects, mostly caterpillars, from the newly opened leaves with their tweezerlike bills. Most of them are fidgety feeders that, as Frank Chapman wrote, "pirouette, or flutter, turning the whole body this way then that, darting or springing here or there, the embodiment of perpetual motion."

The black-capped chickadee appears to use the visible damage done to leaves by caterpillars as a clue in searching for these insects. Bernd Heinrich and Scott Collins did an experiment that showed that captive chickadees were far more likely to search for hidden mealworms (succulent beetle larvae) on birch or maple branches with naturally or artificially damaged leaves than on branches with undamaged leaves. They also reported that some caterpillars eliminate this clue by pruning away damaged leaves. Of 21 caterpillars that are presumably palatable to birds because they are neither spiny nor warningly colored, only one does not cover its tracks by clipping off partially eaten leaves when it has finished a bout of feeding. Presumably unpalatable caterpillars, which are usually warningly colored and are not threatened by birds, do not prune away partially eaten leaves.

The unfurling of leaves in spring, the appearance of the insects that eat them, and the arrival of migrant leaf-gleaning birds that eat the insects are closely synchronized. Trees, insects, and birds act independently of each other, but each, in its own way, heeds the rhythm of the season so that all arrive at the critical stage at the same time.

During the winter, the buds of trees are prevented from growing by a chemical inhibitor that gradually breaks down as the season progresses. Otherwise, the buds might begin to grow in the fall or in response to a few warm days in winter that will inevitably be followed by a return to lethally low temperatures. Once the inhibitor has broken down, their dormancy ends and the buds develop at a rate determined by the accumulating warmth of spring.

The spring emergence of most insects is similarly timed: inhibited until after a long period of cold weather, and then fine-tuned by the increasing warmth of spring. Take, for example, the two species of leaf-eating cankerworms (caterpillars also known as loopers or measuring worms) that are prevalent in the diets of leaf gleaners in spring. The spring and fall cankerworms are named for the season in which the adult moths appear. The eggs of fall cankerworms, laid in late autumn on the bark of trees, are dormant during most of the winter but break dormancy late in that season. Spring cankerworm moths emerge from the pupae in February and March and lay their eggs, which are not dormant, on bark, as do fall cankerworms. The eggs of both species develop slowly and hatch at about the same time in late April and early May, just in time for the newly hatched caterpillars to feed on newly unfurled leaves and to serve as food for migrant leaf gleaners.

As the days of winter and spring lengthen, migratory birds that were content to stay put during the short days of winter undergo physiological and behavioral changes that culminate in migration. Their testes and ovaries increase in size, they accumulate body fat as their appetites increase, and, unlike nonmigratory species, they become increasingly restless as the time to move northward approaches. A bird in this condition is unalterably committed to migrate, but the tempo of its journey and its arrival time at any one place are profoundly influenced by weather patterns and the state of the advancing spring. Most leaf-gleaning migrants keep pace with the unfurling of the leaves as they move northward, but some warblers that are less committed to leaf gleaning than others precede the unfurling

leaves, among them the versatile yellow-rumped warbler, the equally versatile palm warbler, and the pine warbler, which will climb up the trunks or major branches of trees like a creeper as it searches for insects on the bark.

Leaf-gleaning birds, like other birds, have prodigious appetites. In 1905 Ephraim Felt reported that his assistant "saw a pair of tanagers eat 35 newly hatched caterpillars in a minute. They continued eating these minute insects at this rate for 18 minutes; so that . . . they must have eaten in this short time 630 of the little creatures. This would not make them a full meal, as the entire number would hardly be equal in bulk to one full grown caterpillar." At another time, the assistant carefully watched two common yellowthroats as they ate aphids, estimating that they consumed 7,000 within an hour.

As you will read in the next chapter, insects have fought back, evolving various ways to defend themselves against leaf gleaners and other birds that want to eat them. Some insects, including monarch butterflies, make themselves inedible by storing poisons in the tissues of their bodies. But some birds have counterattacked by evolving ways to overcome these defenses. The monarchs that brighten our summer days migrate south to spend the winter in the mountains of the volcanic highlands just west of Mexico City. There they are densely packed—about 4 million per acre—into 12 known sites that range from about a quarter of an acre to about 14 acres in area. Of course, such immense concentrations of insects are generally irresistibly attractive to birds. But of the scores of bird species that occur near these overwintering sites, only a few attack the poisonous monarchs.

Only nine species of birds have been seen to attack monarchs at their overwintering sites, and only two of them, the leaf-gleaning black-headed grosbeak and the black-backed oriole, eat them in quantity. The grosbeaks are able to eat monarchs because they tolerate relatively enormous doses of the poisons that these butterflies contain, doses far greater than those that cause black-backed orioles and most other birds to vomit. The black-backed oriole does not have such a physiological tolerance, but it minimizes its intake of these poisons by eating only small amounts of monarchs with a high concentration of them, and by eating only those parts of a monarch that have the lowest concentration of these poisons. Much as you or I would eat a New England lobster, the orioles strip out and eat the muscles and other tissues of the thorax and abdomen, which are low in poisons, and leave behind the caterpillar's exoskeleton, which contains more of these poisons than any other part of the body.

Even as it is being attacked by leaf-feeding insects, a tree is host to many other insects that bore in its bark and wood, principally beetles and, to a lesser extent, caterpillars and sawflies. The small, white, legless grubs of engraver beetles burrow and feed at the interface of bark and wood, leaving a feather-shaped pattern of tunnels etched partly into the wood and partly into the adjoining surface of the bark—hence the name engraver beetles. In a mass attack these beetles often destroy enough conductive tissue and cambium (growth layer) to kill a tree. Other insects burrow deeper within the tree, eating sapwood and heartwood. Among them are flat-headed borers, white, soft grubs that will become beetles which are often large and handsome in coats of metallic copper, green, or blue; round-headed borers, the soft, white larvae of the large and often gaudily colored long-horned beetles, which are second only to butterflies in their popularity with collectors; carpenterworms, very large wood-boring caterpillars, which as moths have a wingspan of three inches; and, finally, horntails, larvae of certain sawflies, the most primitive of the Hymenoptera (bees, wasps, and their allies). Although the larvae of these insects eat wood, most of them—a few exceptions are now known—cannot synthesize the enzymes that digest cellulose and lignin, the major components of wood, but they do not starve, because they are associated with fungi that provide them with the requisite enzymes.

These and other insects that live in bark or wood are the prey of woodpeckers. All woodpeckers are equipped to dig holes in wood. Their strong, chisel-like bills are efficient woodworking tools; their feet, usually with two toes pointing forward and two pointing backward, clamp them to the tree; and their stiff, spine-tipped tail feathers act as a prop when they climb or hammer away. All of them use their powerful bills to excavate nest cavities into solid wood, but not all of them use their bills to chisel through wood in search of insects. Some woodpeckers seldom or never excavate to find insects, and even those that do also resort to other methods of feeding. Seeds and fruits are a part of the diet of most woodpeckers. I once watched red-bellied woodpeckers gorge themselves on pokeberries in southern Illinois,

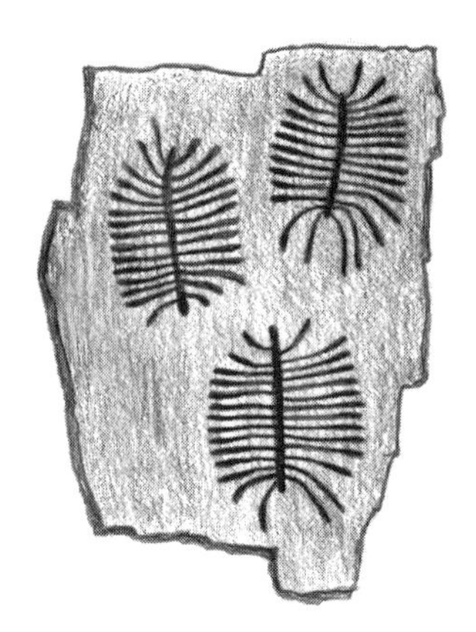

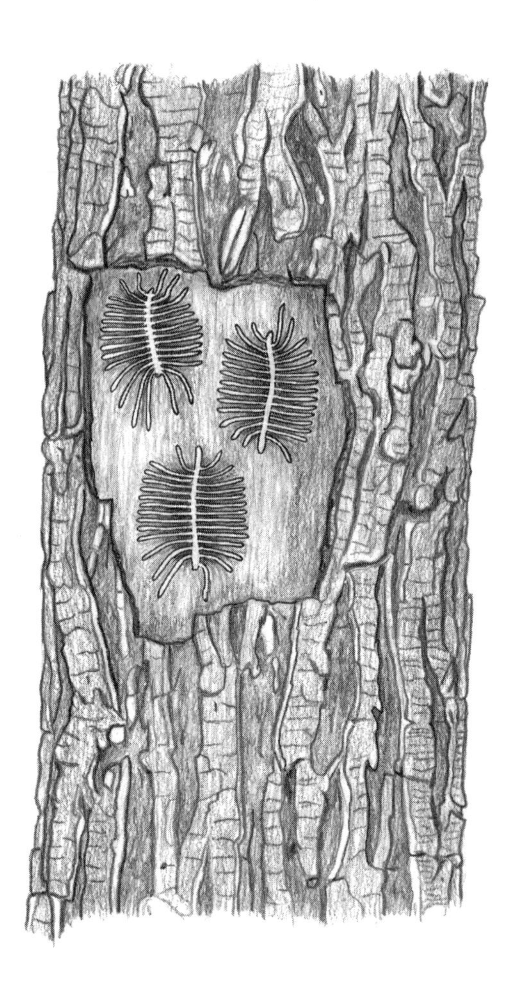

The tunnels made by engraver beetles etched on the wood of a branch and on the inner surface of the corresponding flake of bark

apparently oblivious to their surroundings and so stained by the purple juice of the berries that I did not at first recognize them. Flickers search for ants on the ground, and several woodpeckers, including the acorn, red-headed, and Lewis's, are skillful fly catchers. The woodpeckers known as sapsuckers are unique in their feeding behavior. They peck geometrically spaced rows of small holes into smooth bark, eat some of the soft inner layer of the bark, drink the sweet sap that oozes from the holes, and eat insects that are attracted to the sap. Hummingbirds may be attracted to the sap oozes created by sapsuckers and will drink the sap and eat the associated insects.

The crow-sized pileated woodpecker is the preeminent carpenter among our North American birds. In February of 1925, O. M. Bryens of Three Rivers, Michigan, watched one chiseling into a maple stub, apparently in search of the larvae whose borings Bryens later found in the stub: "After he had left . . . I found some chips near the stub, which were three inches long and one inch wide. Others half this size had been thrown out on the snow a distance of four feet. The hole was on the west side and measured six inches across and ten inches long, and extended to a depth of six inches toward the heart of the stub. There was another hole six inches square on the south side. . . The two holes were dug in about two hours."

In a contribution to Bent's 1939 volume on woodpeckers, Bayard H. Christy reported that, although pileated woodpeckers eat beetle larvae and other wood-boring insects, ants are the chief item in their diet, especially the huge carpenter ants (workers may be as much as half an inch long) that do not eat wood, but do excavate their complex galleries in it. He says that "in pursuit of ants . . . the woodpecker cuts its great furrows in the boles of standing trees, living and dead." He watched a female on the stump of a jack pine flake off fragments of still firm wood, poke her bill into the resulting cavity, and then search for ants with her long, extensible tongue. When I was an undergraduate at the University of Massachusetts in Amherst, I watched a pileated woodpecker as, over a period of several days, it whittled away at the base of a large dead tree, presumably searching for ants. The bird finally chiseled away so much wood that the tree was supported by only a thin shaft of wood. It toppled over in the next strong wind.

The three-toed and the black-backed woodpeckers, inhabitants of our northern forests, are both scalers of bark that gravitate to stands of dead conifers. In *Birds of Massachusetts,* Edward Howe Forbush wrote that the black-backed woodpecker, then known as the arctic three-toed woodpecker, often begins to work at the base of a dead tree.

> It sounds the bark with direct blows, and then, turning its head from side to side, strikes its beak slantingly into and under the bark, and flakes it off. It often works long on the same tree and barks the whole trunk in time, only occasionally working on the branches. Thus it exposes channels of bark-beetles and the holes made by borers . . . In early autumn, while the grubs are still at work on the tree, it lays its head against the

tree, at times, turning it first to one side and then to the other as if listening.

Both the hairy woodpecker and the smaller downy woodpecker excavate trees to find bark- and wood-dwelling insects. But the hairy woodpecker, being heavier-bodied than the downy and having a stouter and longer bill, does so more often and more successfully than the downy. This difference is indicated by a quantitative difference in the kinds of insects found in the stomachs of the two species. According to Bent, the wood-boring round-headed and flat-headed borers and the wood-boring larvae of a few other beetles are the items most often found in the stomachs of hairy woodpeckers, constituting over 31 percent of their annual food intake. But wood-boring larvae make up only about 14 percent of the insects found in the stomachs of downy woodpeckers.

During the winter virtually all insects are inactive and hidden away in the soil, under bark, in the stems of herbaceous plants, in cocoons, or in other protective quarters. At this time of year woodpeckers become opportunists that scrounge widely for any food they can find. I have seen downy woodpeckers enter cornfields to dig overwintering corn borer caterpillars out of the dead stalks. They cling to the dead stems of goldenrods as they peck holes into the inch-wide, spherical galls that house the maggots often used as bait by people who fish through the ice. My co-researchers and I have seen downy woodpeckers cling to 3-inch-long cocoons of the cecropia moth as they pierce the tough, silken, double walls to get at the large and succulent overwintering pupae.

Before we actually saw woodpeckers penetrate cecropia cocoons, Jim Sternburg, my partner in research, and I had collected many cocoons that bore the telltale signs of a woodpecker attack. The two walls of the cocoon and the pupa had been pierced by tiny holes that were often no more than a tenth of an inch in diameter. The viscous, semi-liquid tissues of the pupa had been removed through these holes; and only a few scraps of tissue and the empty skin remained behind. We reasoned that the culprit had to be a woodpecker. Cold-blooded animals were out of the question, because they are not active in the winter, leaving only mammals and birds. Some mammals do prey on cecropia pupae, but none of them have anatomical structures that could pierce a cocoon in this manner. This left only birds. Many of them can pierce a cocoon with their bills, but only woodpeckers have long, extensible tongues with which to extract the contents of the pupa through a tiny hole.

The circumstantial evidence was strong, but, like all scientists, we were not completely convinced until we had direct evidence. We wanted to catch a wild woodpecker in the act or watch a caged woodpecker penetrate a cocoon. At first, we had no luck with wild woodpeckers. We found a few cocoons that had been so recently attacked that icicles of cecropia blood still hung from the hole in the cocoon, but we did not see a wild woodpecker penetrate a cocoon. A year later Jim and I and Aubrey Scarbrough, then our graduate student, did see a wild hairy woodpecker and several wild downy woodpeckers attack cecropia cocoons.

In the meantime, we had asked William G. George of Southern Illinois University in Carbondale to expose some of our cocoons to a caged hairy woodpecker. By the second day after Bill had wired a cocoon to the inside of the woodpecker's cage, the bird had pierced it and hollowed out the pupa. Earlier that day I had called Bill on the telephone and asked him if the woodpecker had as yet attacked the cocoon. He said that it had not. I asked him if he had gone into the cage to take a close look. He answered that he had not, but that he would do so immediately. He soon called back to say that there was a tiny hole in the cocoon with the surrounding silk stained by what he assumed was cecropia blood.

Bill wondered how the woodpecker had known to attack the cocoon. He hypothesized that natural selection has molded woodpeckers, and other birds as well, to be opportunistic, inquisitive explorers of their environment, examining anything new and different to determine if it is edible or contains something edible. To test this idea, he looked at unusual features on trees in a woodland: a healed wound, a bump on a branch, a woody gall, and so forth. On close inspection, he found beak marks, probably those of woodpeckers, on virtually all of these odd features, strong evidence for his hypothesis.

Like woodpeckers, the woodpecker finch and the mangrove finch, the only wood probers on the Galápagos Islands, dig into wood with their strong bills as they search for insects. But lacking the long, barbed tongue of a woodpecker, they use a twig or a cactus spine to probe for insects beyond the reach of their bills. They are among the few tool-using birds in the

world. Like the other Galápagos finches, they are probably the descendants of one species of finch that colonized the Galápagos after these volcanic islands arose from the Pacific Ocean some 5 million years ago. At first devoid of life, the islands were gradually colonized by plants and animals from the distant South American mainland. The newly arrived finch, free of competitors and faced with a plethora of unoccupied ecological niches, underwent an evolutionary adaptive radiation that has, so far, split it into 14 different species. In addition to the two wood-probing species, there are, among others, a warblerlike gleaner, a seed eater with a huge grosbeak-like bill, and a nectar-sipping flower prober. Indeed, Charles Darwin's observations of the Galápagos finches were central to his formulation of the theory of evolution, as has been pointed out by many writers, including David Lack in *Darwin's Finches*.

Far more so than the woodpeckers and Galápagos wood probers, the tyrant flycatchers, a large New World family of 375 species, display diverse foraging styles. Some members of this family feed largely on fruits and some eat fruit only during certain times. The eastern kingbird, for example, eats mostly insects when it nests in North America but eats mostly fruit when it winters in South America. Many flycatchers are air salliers, but some catch prey in other ways. Some tropical species glean foliage, catch prey as they walk on the ground, pick prey off the surface of water, or, like the great kiskadee, dive into water to catch fish or tadpoles. Many of the tropical species and all but 2 of the 30 or more species that nest in North America are true members of the air-sallying guild. They fly out from a perch to snatch an insect or spider from the air. The two exceptions are the kiskadee, which ranges as far north as the Rio Grande Valley of Texas, and the northern beardless tyrannulet, mainly a foliage gleaner.

Air-sallying flycatchers will eat almost anything that flies, and also spiders that "balloon" through the air on long strands of silk. Their insect food includes many species from many different families of all the major orders of insects: damselflies, dragonflies, grasshoppers, true bugs, aphids and their allies, lacewings, beetles, moths and butterflies, flies, and wasps and bees. I said *almost* anything. Flycatchers, like all birds and other ani-

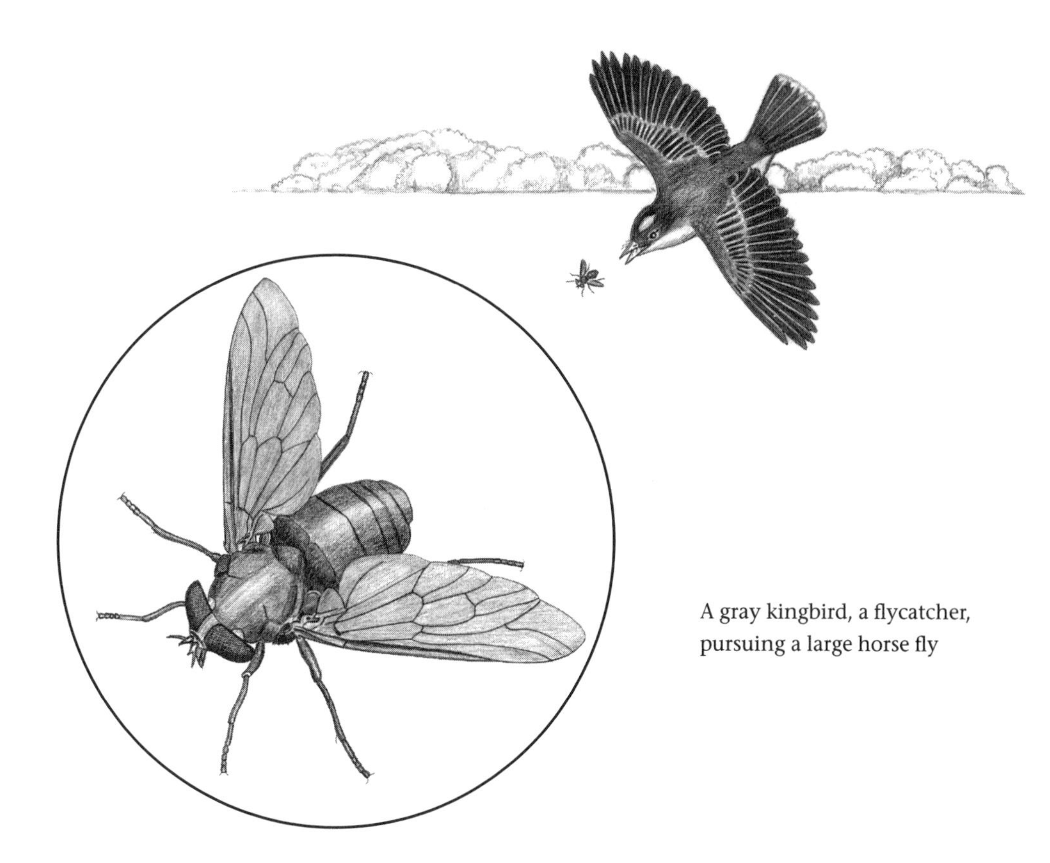

A gray kingbird, a flycatcher, pursuing a large horse fly

mals, sometimes reject a prey item, perhaps because they know that it can sting or make them sick. Black phoebes, for example, are known to pass up certain butterflies that fly past within a few feet of their perch, not even sallying out to examine them more closely.

Some flycatchers and a few other air salliers are able to eat wasps and bees, because they have evolved ways to avoid being stung or because they recognize and attack only the stingless males. Among them is the brown-crested flycatcher of the American West, which is often seen hawking bees from a perch near an apiary. In the belief that this flycatcher is a serious pest that eats worker bees, bee keepers often refer to it as the bee martin. In *Arizona and Its Bird Life,* Herbert Brandt quoted a perceptive Arizona bee

keeper who questioned the supposed destructiveness of the brown-crested flycatcher: "after examining the stomach contents of a large number [of brown-crested flycatchers], during a period of more than 20 years, and not in a single instance finding the remains of a worker bee, nor finding a bee-sting in the mouth or throat of one of these birds, I became convinced that it did not prey on worker bees, but only on drones (the stingless males)." According to Bent, other flycatchers also recognize drones and attack them rather than the far more numerous stinging workers. Of the 50 honey bees found in the stomachs of several eastern kingbirds, 40 were definitely drones, 6 were smashed beyond recognition, and only 4 were workers; 31 honey bees, 29 of them drones, were found in the stomachs of 5 western kingbirds.

The same perceptive bee keeper observed that summer tanagers also ate honey bees. They perched in the trees that shaded the hives and swooped down on incoming bees as they slowed down to alight at the hive entrance. Reporting on the feeding behavior of the summer tanager, the bee keeper wrote that "its food in the areas about my apiaries consisted almost entirely of bees, and worker bees, at that. Or I would better say, parts of bees, for the bird skillfully avoided contact with the stinger end of its victim by breaking off that end. This was accomplished by catching the bee across the middle of the body and, on alighting upon a branch or other perch, breaking off the protruding end of the abdomen by giving it a sharp sweep across the perch." He went on to note that "the top of nearly every hive was sprinkled with the abdomens of bees, and just as many probably fell on the uncovered ground." The western tanager also ate worker bees, but its depredations were less serious because it was much less abundant in the apiary.

You might guess that hunting for insects on the bark of a tree would be an unprofitable enterprise for a bird. But you would be wrong. There are more insects in the bark habitat than meet the eye at first glance, and anywhere in the world where trees grow, there are birds that make their living largely by gleaning insects from bark. Their prey are the many different kinds of insects and other arthropods that can be found under loose scales of bark, in the crannies and interstices of bark, or resting or crawling on its surface. Preeminent among North American bark gleaners are the brown creeper and our four species of nuthatches. But there are other birds that are not as strongly committed to this method of hunting but that sometimes use it to obtain a significant portion of their insect diet. Among them are woodpeckers, chickadees, titmice, Bewick's wren, Scott's oriole, and several warblers—including especially the black-and-white, pine, yellow-throated, Grace's and olive warblers.

Arthropods differ in how closely they are associated with the bark habitat. Some spend their entire lives there, others are there during only some of their life stages, and some are only visitors—but all of them are fair game for the bark gleaners. Some of the visiting insects have no regular association with bark but light on its surface by happenstance and remain there only briefly. But other visitors have a consistent association with bark. Among the latter are many different kinds of nocturnal moths that regularly spend the daylight hours resting motionless on the trunk or a large branch of a tree. As you will find when you read the next chapter, they are so beautifully camouflaged that they are virtually invisible when they sit on bark. Anyone who has ever tried to collect these insects knows that they often escape the notice of people who are searching for them, and experimental evidence shows that they often escape the notice of insect-eating birds as well.

Many insects that feed on the foliage of trees, among them the two species of cankerworms, are exposed to bark gleaners at certain times during their life. On emerging from a pupa in the soil, a female cankerworm moth of either species, wingless as are the females of several other species of moths, crawls up the trunk of a tree, mates with a winged male, and lays her eggs in clusters. Males rest on bark during the day. The spring cankerworm lays her eggs under loose scales of bark on the trunk or a main branch, and the fall cankerworm lays them on the surface of a twig or a small branch. The newly hatched caterpillars crawl over the bark to the foliage, and the fully fed caterpillars, ready to pupate, make their way to the ground by crawling down the tree or by dropping from the foliage on a strand of silk. Cankerworms are thus exposed to bark-gleaning birds at the beginning and end of their caterpillar stage and throughout their egg and adult stages, but not during the pupal stage and most of the caterpillar stage.

The thick, corky bark of a tree may house a miniature ecosystem—founded on tiny plants and wind-blown organic matter—that includes as

permanent residents herbivores, scavengers, and predators, and that is regularly invaded by bark-gleaning birds, ants, and other predaceous insects. Among the permanent residents are various species—all very small—of mites, springtails, and barklice that live in cracks and crevices and eat lichens, algae, mold spores, pollen grains, and any other organic matter they can find. They are accompanied by parasites and by the predators that eat them, including spiders, predaceous mites, beetles, and pseudoscorpions (tiny, harmless arachnids that cannot sting but are vaguely similar in appearance to true scorpions). All of these insects—herbivores, scavengers, and predators—are food for bark gleaners.

The surface of smooth, thin-skinned bark is often occupied by many insects and other arthropods that take their food from the tree itself. Among them are various species of Homoptera that live by piercing the thin bark and sucking sap. The pine bark aphid, for example, covers itself with a wooly layer of wax, and, as F. C. Craighead wrote, a large, heavily infested tree appears to have been coated with whitewash. Living in these colonies may be insects that eat the aphids, such as adults and larvae of lacewings and certain ladybird beetles. Many kinds of scale insects, so-called because they secrete a waxy scale that covers their body, also live on the bark of trees and suck sap. Among them is the oyster shell scale, which attacks many different kinds of forest, shade, and fruit trees and is commonly found in southern Canada and the northern two thirds of the United States.

Oyster shell scales survive the winter as eggs, as many as 150 of them under a waxen scale that once housed their mother. Late in spring, tiny nymphs with functional antennae, mouthparts, and legs—called crawlers—hatch and move about as they search for a suitable place to settle down, sometimes on their home tree and sometimes on another. Only in this stage are both males and females mobile, and only in this stage can these scales colonize new trees by moving from one tree to another, as when a female crawler is blown by the wind or carried on the body of a bird or an insect. The crawler ultimately settles down, inserts its mouthparts into the living tissue of the bark, secretes a covering scale, and molts its skin. With this first molt, both males and females lose legs and antennae, becoming little more than sacs with mouthparts on one end and an anus on the other. The scale insect, essentially a parasite of the tree, remains in this degenerate condition as it sucks the sap of the tree, grows, and undergoes several more molts until it is finally ready to become an adult. Adult females remain under the scale and are little changed, still eyeless, wingless, and legless. The males, however, emerge from under the scale with functional eyes, antennae, wings, and legs. They find a mate by homing in on a sex-attractant pheromone that females release into the air.

Each species of bird that eats bark-dwelling insects adds its own characteristic twist to the business of bark gleaning. Brown creepers hunt mainly on tree trunks, starting near ground level and climbing upward as they spiral around the trunk, propping themselves with their woodpecker-like tails as they methodically search cracks and crannies for spiders, ants, beetles, mites, pseudoscorpions, insect eggs, and other prey. When a creeper reaches the first main branches, it usually glides through the air to a nearby tree to continue foraging, using the energy-saving maneuver of planing down to the base of the other tree and again climbing upward as it forages. Nuthatches generally forage on the trunks and main branches, sometimes climbing upward and sometimes climbing downward head first. Like creepers, they search for insects in cracks and crevices but often also pry up loose scales of bark. Douglas H. Morse reported in 1968 that, in Tangipahoa Parish in Louisiana, brown-headed nuthatches used small flakes of bark held in their bills as levers or wedges to pry larger scales of bark from longleaf pines. Black-and-white warblers clamber quickly over trunks and main branches, moving upward, downward, and even spirally. In the spring of 1996 in central Illinois I watched two pine warblers crawl slowly and deliberately straight up the trunk of a white oak as they painstakingly searched for insects. In the South, yellow-throated warblers are found in pines or cypresses, but in the Mississippi Valley they prefer to live in sycamores along the banks of streams. They forage on the trunks and main branches of these trees, crawling both upward and downward much like black-and-white warblers.

In 1926, a year before Charles Lindbergh made the first flight across the Atlantic, entomologists at Tallulah, Louisiana, were chasing insects with an airplane, a biplane with insect traps mounted below its upper wing. An

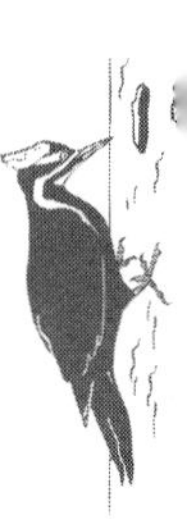

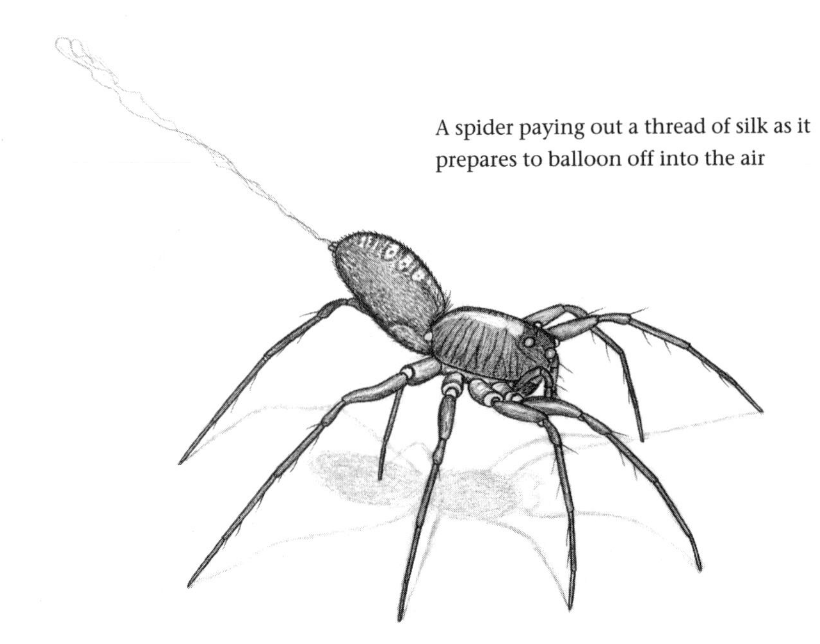
A spider paying out a thread of silk as it prepares to balloon off into the air

article by Perry Glick, complete with photographs, tells the story in fascinating detail. From his data, I calculated that during daylight in May, a volume of air 1 mile square and extending from 20 feet above the ground to an altitude of 500 feet contained over 32 million arthropods. This amounts to 6 arthropods per 10 cubic yards of air, remarkably close, considering the many variables involved, to the 8 arthropods per 10 cubic yards found by Matthew Greenstone and his co-researchers in July of 1990 near Bellflower, Missouri. Ten cubic yards is quite a small space, about the size of a small clothes closet. The abundance of airborne arthropods decreases with increasing altitude but increases rapidly as ground level is approached. Some species are present at night, including many not seen during the day, but they are only about half as numerous as the day fliers.

Mites, spiders, and 18 orders of insects were represented in the samples taken at Tallulah, including members of 216 families, 824 previously known genera, 4 new genera, at least 700 previously known species, and 24 new species. Flies were by far the most abundant, and beetles followed as a distant second. Most of these insects were winged, but there were also some wingless ones and larvae that, like the mites and spiders, were passively swept along by air currents. Spiders, which balloon on the breeze as they hang from long threads of silk, were numerous, and one was captured 15,000 feet above the ground.

This abundance of arthropods in the air, often referred to as aerial plankton, is a rich resource that is exploited by birds, bats, and dragonflies. Swallows, swifts, and dragonflies pursue these arthropods during the day, and bats, nightjars, and nighthawks take over the night shift. The swallows, swifts, and nighthawks have in common long, falcate (sickle-shaped) wings and wide, gaping mouths that open almost back to the eyes, adaptations that facilitate their habit of sweeping insects out of the air. These similarities came about through convergent evolution, similar demands by the environment evoking the evolution of similar characteristics in unrelated organisms. As Elliot Coues, a nineteenth-century physician and ornithologist, put it, these similarities are a matter of "analogy, not affinity."

Swallows sometimes hunt well above the ground, but also swoop down to take insects from just above the surface of a pond or just above the ground. Like the cattle egrets and cowbirds that you will meet shortly, barn swallows take advantage of disturbances that flush insects from the vegetation. As Forbush wrote of this bird in 1907,

> It follows the cattle afield or swoops about the house dog as he rushes through the tall grass, and gathers up the flying insects disturbed by his clumsy progress. When the mowing machine takes the field, there is a continual rush of flashing wings over the rattling cutter-bar just where the grass is trembling to its fall. The Barn Swallow delights to follow everybody and everything that stirs up flying insects—even the rush and roar of that modern juggernaut, the motor car, has no terrors for it.

Swallows, especially purple martins, have been favorites of Americans since before the continent was invaded by Europeans, in part because of their reputation as hunters of mosquitoes. Forbush wrote of Native Americans hanging hollowed-out gourds near their homes as nesting places for martins. People still put up martin houses throughout much of southern Canada and the United States. In 1946, Aretas A. Saunders, a well-known ornithologist of his day and, by great good fortune, my high school biology teacher, took me and my friend Robert Braun to see one of the few remain-

ing martin colonies in Connecticut. It was on a large estate in Greens Farms, and survived only because the caretaker shot house sparrows to keep them from monopolizing the martin houses. The purple martin is still no more than a rare nester in Connecticut, largely because of competition for nesting sites from house sparrows and starlings. It was much easier to attract nesting purple martins before the pestiferous house sparrows and starlings were introduced from Europe.

Swifts are the most aerial of birds. They feed, drink, gather twigs for nests, bathe, court, and often mate while flying. Some may even spend the night on the wing. The chimney swifts of the eastern United States and Canada usually forage higher in the air than do swallows, often well above the rooftops or trees and sometimes much higher than that. They eat no vegetable food and subsist almost wholly on flying insects, including many species of flies, beetles, wasps, winged ants, bees, true bugs, and Homoptera. When feeding their nestlings, swifts bring to the nest a food bolus carried in the mouth and consisting of hundreds of small insects held together with the same sticky saliva that they use to glue their nest to the wall of a chimney.

The common names of many North American nightjars—poor-will, whip-poor-will, chuck-will's-widow—are onomatopoeic renderings of their calls, probably reflecting the fact that we rarely see these nocturnal birds but often hear their loud calls at night. They are the night shift of the insect-eating aerialists. Well-equipped to catch insects in the air, their heads seem to be all mouth, and their yawning gape is widened by a surround of stiff bristles. The three nightjars mentioned above often fly close to the ground as they feed. They eat almost any insect that flies at night but seem to be especially fond of large moths. Whip-poor-wills eat such giants as the cecropia, luna, and polyphemus moths. Chuck-will's-widows, largest of our North American nightjars, feed mainly on insects, but small birds such as hummingbirds, swallows, sparrows, and warblers have been found swallowed whole in their stomachs.

Like the chimney swift, the common nighthawk, which likes to nest on gravel roofs, has become a resident of cities and towns. The nighthawks, three species of which occur in North America, are not as strictly nocturnal as the nightjars. They rest during the middle of the day, but feed in the morning and late afternoon as well as at night. Common nighthawks

A whip-poor-will pursuing a cecropia moth

swoop high over the rooftops as they pursue all sorts of insects from mosquitoes to large moths. According to Bent, the stomachs of some nighthawks have contained no less than 50 species of insects.

On a fine day in early spring, a big, green John Deere tractor pulls a gangplow through a bottomland field in eastern Missouri. A flock of cattle egrets follows behind, walking the newly turned earth as they harvest the many different kinds of soil-dwelling insects that the plow exposes. Among the insects eaten by the egrets are fat cutworms that would soon have transformed to moths; long, thin wireworms, the larval stage of click beetles; and C-shaped white grubs that would have become May beetles. On another day or in another place in North America, other birds, perhaps grackles, crows, cowbirds, or Franklin's gulls, might be seen following another plow in another field.

Cattle egrets made their way across the Atlantic from Africa to South America, apparently without the aid of humans but probably assisted by the northeast trade winds. They were first seen in Guyana, on the northeast coast of South America, in the late nineteenth century, and since then they have spread throughout much of South and Central America, the islands of the Caribbean, and North America. In 1952, seven pairs were found breeding in Florida. Ten years later this small group had grown to a colony of about 4,000 birds. Today, cattle egrets are among the most abundant herons in North America, and nest in southern Ontario and most of the continental United States. Wherever they occur, they associate with cattle and agricultural machinery.

For our immigrant cattle egrets, plows and domestic cattle are substitutes for the elephants, rhinoceroses, and other large animals with which cattle egrets back in Africa still associate. As these large animals move about, they inadvertently act as beaters that flush the insects and other small creatures on which the egrets feed. Associating with beaters greatly increases the feeding efficiency of these birds, as was shown by Harold Heatwold in Puerto Rico. By simultaneously observing two cattle egrets in the same pasture, one associated with a cow and one foraging alone, Heatwold and another observer determined that, while taking only two thirds as many steps, egrets associated with cattle captured from 25 to 50 percent more prey than did egrets foraging alone. Similarly, A. L. Rand found that the foraging success of groove-billed anis in El Salvador was increased when they associated with cattle. In the dry season, when insects were scarce, an ani hunting alone found only one third as many insects as it did when it accompanied a cow.

Cattle egrets, Heatwold discovered, were much more likely to associate with cows that were actively grazing in the sun than with cows that were in the shade or engaged in other activities. They stayed close to the grazing cow, within about 3 feet, and usually fed near its muzzle or front legs. An examination of 20 stomachs revealed that the cattle egrets ate mainly grasshoppers, crickets, beetles, flies, lizards, and frogs. All are creatures likely to be flushed by grazing cattle.

For some native North American birds, among them crows, cowbirds, and Franklin's gulls, plows and domestic cattle are surrogates for the great herds of bison that once wandered the plains. Today, the herds are gone, and birds that had associated with them for many millennia, since long before humans invented agriculture, now associate with cattle, plows, and various other farm implements that flush insects from hiding as the bison once did.

In his 1929 monograph on the cowbirds, Herbert Friedmann, then a professor at Amherst College in Massachusetts, wrote that brown-headed cowbirds were—even at that late date—still known as buffalo birds in some parts of the West. He also quoted from early accounts of the association of these birds with bison. One traveler on the plains reported that in 1819 and 1820 "cow buntings" (brown-headed cowbirds) were often seen "flying and alighting in considerable numbers on the backs of the bisons, which from their submission to the pressure of numbers of them, seem to appreciate the services they render by scratching and divesting them of the vermin."

Other early observers also assumed that cowbirds associate with grazing animals in order to eat ticks or other parasites that infest their bodies, flies that swarm about them, or intestinal worms that are discharged with their droppings. But F. E. L. Beal's examination of the stomach contents of brown-headed cowbirds does not bear out this assumption. No intestinal worms were found, and insects that annoy bison and cattle constituted only a very small fraction of what had been eaten. In actuality, grasshoppers and leafhoppers, insects that live hidden in the grass, are the favorite animal food of brown-headed cowbirds.

There are a few birds, sometimes called cleaners, that do associate with other animals to eat ticks and other parasites that they pluck from their bodies. Two African birds, the red-billed and yellow-billed oxpeckers, associate with cattle or large wild animals such as rhinoceroses, cape buffalo, hippopotamuses, and giraffes. They spend most of their time sitting or roaming on their host's body as they pluck ticks, their major food, from the skin, but they occasionally fly out to catch an insect in the air. In the Galápagos Islands, the native mockingbird occasionally visits land iguanas to obtain ticks. As Keith Christian put it, when a mockingbird landed on its back, the iguana "assumed a cooperative posture . . ., raising itself off the ground as high as possible on all four legs and remaining motionless while the mockingbird picked ticks off its body."

But Herbert Friedmann understood the role of large animals as beaters that flush insects for their avian companions. He wrote of the brown-headed cowbird:

> In grazing, the animals scare up the swarms of grass-inhabiting insects, especially grasshoppers and locusts, with every bite, and in this way render the food supply, which was already there but hidden, far more available to the birds. It is chiefly during the summer and fall that the Cowbirds gather around their four-footed companions and it is in this same season that the grasshoppers are most abundant.

One bird even uses another as a beater. The carmine bee-eater of East Africa, 14 inches long if you count the elongated central tail feathers, uses the kori bustard, as much as 40 inches tall, as a beater and a moving perch from which to hunt its prey. As the larger bird moves over the grassy plain in search of its food, which consists principally of large insects, particularly grasshoppers, the bee eater swoops down from its back to snap up the grasshoppers and other large insects that the bustard flushes. Since the bee eater eats the bustard's food, it is actually a kleptoparasite: it steals from the bustard.

Over 200 species of birds are kleptoparasites. The most famous of them are the skuas and jaegers, marine birds that harass gulls and other victims to force them to drop or disgorge food. Besides the carmine bee eater, a few other birds that eat insects or other small creatures occasionally steal food from other animals. Starlings and brown thrashers have been seen snatching earthworms away from American robins. Several different kinds of hummingbirds steal insects from spider webs. House sparrows and American robins will lie in wait to take grasshoppers from passing digger wasps that are in the process of stocking their underground nests.

Just as cattle egrets and cowbirds increase their catch of insects by following cows or farm machinery, some South and Central American birds accomplish the same end by following army ants of the species *Eciton burchelli,* fierce predators that hunt insects and other small animals as they relentlessly advance through the forest in dense swarms that may be as much as 60 feet wide. T. C. Schneirla's description of the sounds that emanate from a raiding swarm vividly expresses its size and its overwhelming effect on the animals in its path.

> The approach of the massive *burchelli* attack is heralded by three types of sound effect from very different sources. There is a kind of foundation noise from the rattling and rustling of leaves and vegetation as the ants seethe along and a screen of agitated small life is flushed out. This fuses with related sounds such as an irregular staccato produced in the random movements of jumping insects knocking against leaves and wood. This noise, more or less continuous, beats on the ears of an observer until it acquires a distinctive meaning almost as the collective death rattle of the countless victims. When this composite sound is muffled after a rain, as the swarm moves through soaked and heavily dripping vegetation, there is an uncanny effect of inappropriate silence.
>
> Another characteristic accompaniment of the swarm raid is the loud and variable buzzing of the scattered crowd of flies of various species, some types hovering, circling, or darting just ahead of the advancing fringe of the swarm, others over the swarm itself or over the fan of columns behind.

A raiding swarm flushes small vertebrates and a host of insects and other arthropods, among them spiders, scorpions, beetles, cockroaches, grasshoppers, ants of other species, and a great variety of other forest insects. Most of this booty is harvested by the ants, but some of it is appropriated by various species of birds that follow the ants, not to eat the ants, but to pirate a share of the multitude of insects and other small living things that the ants so efficiently flush from hiding.

On Barro Colorado Island in Panama, at least ten species of birds accompany swarms of army ants. Their calls are characteristic sounds of the forest. As Schneirla wrote:

> No part of the more prosaic clatter, but impressive solo effects, are the occasional calls of antbirds. One first catches from a distance the beauti-

> ful crescendo of the bicolored antbird, then closer to the scene of action the characteristic low twittering notes of the antwren and other frequenters of the raid. For locating swarm raids these are most useful clues as a rule, since the birds ordinarily are to be heard at or near the scene of action from the time of first morning light when the raid begins.

Many different species of birds more or less regularly associate with swarm-raiding army ants. Among them are various kinds of cuckoos, puffbirds, motmots, woodcreepers, tanagers, and some, but not all, species of the antbird family (Formicariidae, from the Latin *formica,* ant) that have been given names such as antthrush, antshrike, and antwren. At any one place and at any one time, a swarm may be followed by as many as 50 birds of 20 or more species.

These birds fight and jockey for positions near the leading edge of the ant swarm, but, in the main, they divide the booty among themselves by taking up different positions with respect to the swarm and by using different techniques to catch their prey. Some hunt from the ground, some from low perches on shrubs, some woodcreepers from tree trunks as they cling like woodpeckers, and some, notably puffbirds and motmots, from higher perches in the trees. Most of these birds dart forth to snatch insects from the ground or from the air. But puffbirds and motmots wait quietly for the ants to ascend the trees and flush grasshoppers and scorpions from arboreal tangles of vines and other vegetation. Then these birds fly out to seize the prey from a branch or a leaf and take it back to their perch for the kill.

The army ant–bird association is even more complex than Schneirla's description indicates. The birds that are attracted to ant swarms are in themselves an attraction to certain insects. In Costa Rica, as reported by Thomas Ray and Catherine Andrews, large numbers of ithomiid butterflies orient to swarms of army ants, possibly homing in on odors, to suck liquid from the birds' droppings, liquid that contains proteins and other nutrients that the butterflies require to produce eggs. At any given moment, many of the butterflies will be feeding on the bird droppings and others will be flying low over the ground searching for them, stopping to examine white spots such as fungi and lichens that resemble droppings. These butterflies, which are toxic and warningly colored, are not attacked by the birds that follow the ant swarm.

African and Asian birds known as honeyguides eat insects of various kinds but, unlike almost all other animals, they are also greedy consumers of the wax that honey bees and some other insects secrete. In a fascinating treatise on the honeyguides, Herbert Friedmann quotes an early reference to these birds. In 1569 João dos Santos, a Portugese missionary to east Africa, mentioned the *"sazu passaro que come cera,"* which Friedmann translated as "*sazu,* a bird that eats wax." (*Sazu* is the local African name for two species of honeyguides.) As Friedmann put it, "It appears that this bird attracted the attention of the good padre because he noted it flying in through the open window into his mission church to feed on the bits of wax in the candlesticks on the altar." In nature, most honeyguides eat the waxen combs produced by honey bees, but a few species eat the wax that covers the bodies of some scale insects.

Only a few animals can digest and assimilate wax, among them the yellow-rumped warblers and tree swallows that I discussed in Chapter 2. There is no doubt that honeyguides also digest wax; at least one species can live for months on a diet of nothing else. But we do not yet know how these birds manage to digest it. Is the wax actually digested by friendly bacteria that live in the intestines of the bird, or are honeyguides among the few animals that can themselves make an enzyme that breaks wax down into digestible fats?

One of these wax-eating African birds, the greater honeyguide *(Indicator indicator)* mentioned in Chapter 1, leads people or other animals, usually honeybadgers but sometimes baboons, to the nests of honey bees in order to eat the combs, honey, and bee larvae that are left behind after the larger animal has broken into the nest and eaten its fill. With the probable exception of the poorly known scaly-throated honeyguide, the other so-called honeyguides are not really guides. They do not seek help from other animals. They eat bee's wax where and when they can find it, sometimes from combs that protrude from a bee nest and sometimes by entering nests with large entrances.

Friedmann described the typical guiding behavior of the greater honey-guide. The bird approaches a person and calls attention to itself by fluttering about close to the person or by making an elaborate display as it sits on a conspicuous nearby perch. In either case, it constantly sounds a loud, repetitive, churring call. A perched bird fans its tail to reveal its white outer tail feathers and arches and ruffles its wings so as to display its yellow "shoulder" bands. If the person comes within 15 to 50 feet, the bird flies off with a downward swoop, its outer tail feathers widely spread, and goes off to a more distant perch. There it waits, all the time churring loudly, until the follower again comes near. Then the display is repeated. This goes on until the bird reaches the vicinity of the bee's nest, at which point it stops calling and perches quietly in a nearby tree until its follower has opened the nest and departed with its loot. Only then does the bird come down to feed on the leftovers. In some parts of Africa, people still follow honeyguides in order to obtain honey, but, according to Friedmann, this custom has been declining as modern ways encroach on native ways.

African legends have it that you must leave something behind for the honeyguide. Otherwise it may not lead you again in the future, may take its revenge by leading you to a lion or some other dangerous animal the next time, or may come and chatter at you when you are stalking game and thus reveal your presence to your intended quarry.

# The Bugs Fight Back

# 4

Just as birds have evolved tactics for capturing insects, insects have, in response, evolved ways to protect themselves against birds, their foremost enemies among the vertebrates. Many insects, probably the majority, are camouflaged—often so deceptively disguised that insect-eating birds pass them by. Some of these camouflaged insects will, if they are discovered, startle and confuse an attacking bird by suddenly revealing brightly conspicuous "flash colors." Birds generally pay no attention to the various insects that resemble some familiar but inedible object, perhaps a thorn, a twig, or even a bird dropping. Others insects frighten off attacking birds by their resemblance to a bird-eating predator, usually one that is larger than the insect itself. This is not as improbable as it may seem. A large caterpillar can pass for the head and neck of a small snake; a large moth with wings spread wide can look like the staring face of an owl. Insects that sting, discharge noxious chemicals, or are toxic if eaten advertise their defensive capability by conspicuous behaviors and color patterns that birds and other insect eaters can easily learn to heed. Some defenseless insects may manage to ward off insect-eating birds by bluffing, by mimicking a toxic or otherwise well-protected insect. None of these defensive tactics is infallible, but they do improve the odds for survival.

One of the ways in which an insect may avoid becoming a meal for a bird was graphically described in a 1924 letter from Colonel A. Newnham, who was stationed in India, to Professor E. B. Poulton of Oxford University:

> I came across the larva in question in the month of August or September 1892, at Ahmadabad on a bush of Salvadora . . . I was stretching across to collect a beetle and in withdrawing my hand nearly touched what I took to be the disgusting excreta of a crow. Then to my astonishment I saw it

> was a caterpillar half-hanging, half-lying limply down a leaf. [Something] that struck me was the skill with which the colouring rendered the varying surfaces, the dried portion at the top, then the main portion, moist viscid, soft, and the glistening globule at the end. A skilled artist working with all materials at his command could not have done it better.

Other insects and even some spiders also look like bird droppings. Among them are a number of North American insects, including the caterpillars of some of our common swallowtail butterflies. The caterpillars of an African bombycid moth look like bird droppings, and like Colonel Newnham's caterpillar, sit on the upper side of the leaves of their food plant. When they are small they are gregarious, and a group of them gives the impression of a number of small feces dropped by several small birds that had been roosting above. When they grow large, they are solitary and look like a single dropping of a large bird or lizard. In *A Naturalist in the Guiana Forest,* Major R. W. G. Hingston, leader of the 1929 Oxford University expedition to British Guiana, described a small adult moth that resembles a flattened bird dropping. Its wings are glossy white, almost pearly, and have some patches of dark brown and areas with a blush of slaty gray and traces of yellow. This moth, which habitually rests on the upper side of a leaf with its wings spread out and flattened down against the surface, looks like a small bird dropping that fell from a height and smashed itself against the leaf.

Some insects are almost indistinguishable from thorns, twigs, or even small clumps of soil. An east African planthopper has a remarkable way of passing itself off as a flower. Each individual, about a fifth of an inch long, looks like a perfect little blossom. But when these planthoppers gather together as a resting group, they look like an inflorescence composed of many closely spaced blossoms, and the deception is virtually perfect. When they line up head to tail along a vertical stem, all with their heads pointing up, they look for all the world like a spike of lupine blossoms. As Wolfgang Wickler wrote, even botanists have been deceived, plucking an inflorescence only to have all the blossoms fly away.

✻ How did such precise resemblances ever come to be? The answer is that organic evolution, proceeding over many millennia and countless gen-

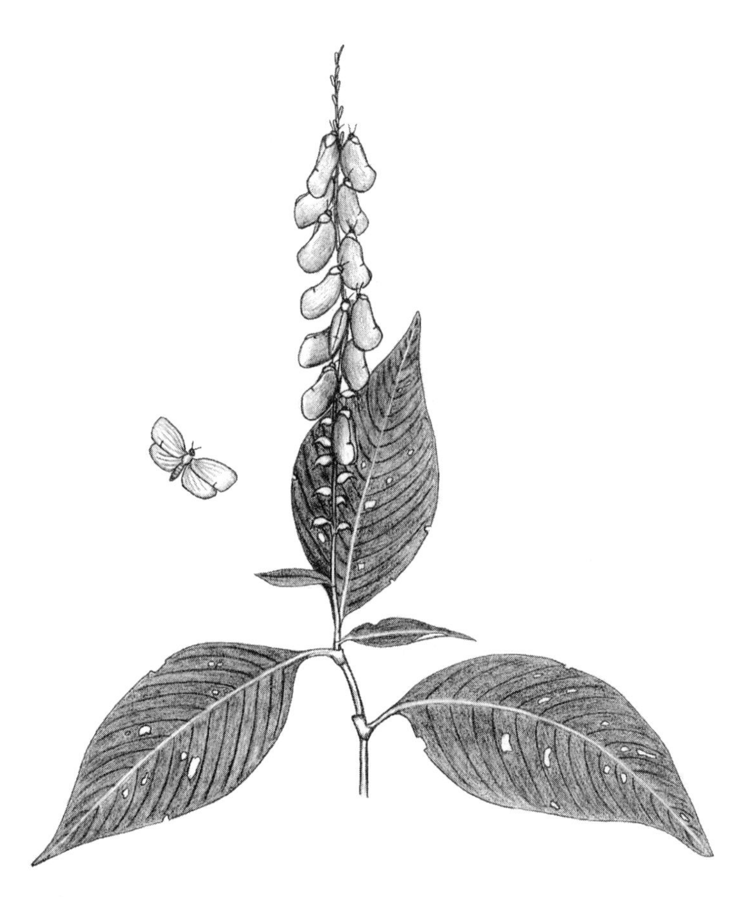

A flying planthopper coming to join other planthoppers already lined up on a stem, where they look like a spike of flowers

erations, produced caterpillars that look like bird droppings and planthoppers that look like flowers—as well as all the other biological wonders that we see around us. The essence of the evolutionary process is natural selection. Charles Darwin had the brilliant insight that natural selection produces new species in essentially the same way that dog fanciers produce new breeds by artificial selection, by allowing only dogs with heritable, desirable characteristics to become the parents of the next generation of dogs. Similarly, natural selection tends to weed out the less fit individuals, and to permit the survival of those best adapted to cope with the hazards of their environment or to take advantage of the opportunities it offers. A

well-camouflaged individual, for example, is not as likely to be eaten by a predator as is a poorly camouflaged one; a moth with a long tongue can reach the nectar in flowers that are too deep for a moth with a shorter tongue.

Generally speaking, the fittest individuals, those that successfully run the gauntlet of natural selection, will be the parents of the next generation, and as natural selection continues to operate generation after generation, their useful new heritable traits will spread throughout an interbreeding population. If this population becomes separated from other populations of its species, most often geographically but sometimes by a switch to a different habitat or host plant, it is likely to become a new species as new, favorable heritable traits accumulate. If some of these newly evolved traits are courtship behaviors that preclude mating with other populations, natural selection will have produced a different species that "breeds true" and is reproductively isolated from all other species.

But where do new, useful, heritable traits come from? This question greatly troubled Darwin. Not until after his death was the new science of genetics founded, the science that would ultimately answer this crucial question. We now know that new traits that can be inherited are constantly appearing through the mutation of DNA, the stuff of the genes that determines the anatomy, physiology, and behavior of all living things. Mutations are caused by environmental factors such as ultraviolet light and cosmic rays and, as molecular biologists have recently shown, by intrinsic factors in the DNA itself. Useful mutations tend to survive and be passed on, but the many deleterious mutations are ultimately eliminated by natural selection.

We generally think of natural selection as a process so glacially slow that evolutionary changes are imperceptible within a human lifetime. But natural selection sometimes proceeds quite rapidly and has been observed in action several times. The most widely known and one of the most thoroughly understood examples, that of the exquisitely camouflaged peppered moth, was elucidated by H. B. D. Kettlewell of Oxford University. In the eighteenth century and presumably before that, peppered moths were always or almost always light in color and were therefore inconspicuous on the light-colored bark of trees, on which they rested during the daylight hours. Their camouflage was enhanced by markings on their wings and bodies that resemble lichens that grow on the bark of trees. But the forests in which these moths live changed during the Industrial Revolution. As coal-burning factories multiplied, beginning in the early nineteenth century, smoke pollution increased. Tree trunks in woodlands near factory towns were stained black with soot, and the lichens that grew on their bark died. The light-colored moths were no longer camouflaged when they sat on these dark tree trunks. In fact, they were perilously conspicuous. Only black peppered moths would have gone unnoticed on the smoke-stained bark. They were almost unknown in the eighteenth century; the occasional black individual that appeared was probably soon snapped up by a bird because it was conspicuous on light-colored bark.

A black peppered moth was first noticed in a previously all-light population in 1848, in Manchester. The population of the better camouflaged black form increased rapidly as the population of the light form decreased. By 1898, only 50 years later, 95 percent of the peppered moths in the smoke-polluted woodlands near Manchester were black. The same thing happened to this moth and other insects in woodlands near factory towns all over Great Britain, northern Europe, and, to other insect species, in North America. Although the rare black mutants probably appeared occasionally in unpolluted places, natural selection eliminated them, and in unpolluted areas, peppered moths continued to be light-colored. The story of the peppered moth has an interesting sequel that illustrates the ceaseless operation of natural selection. As Britain cleans up air pollution, the trunks of stained trees revert to their original light color and lichens once again grow on their bark, favoring the small minority of light-colored moths that persisted in smoke polluted areas. Where air pollution is decreasing, light-colored moths are supplanting black moths.

But what is the selective agent that eliminates poorly camouflaged peppered moths? Birds that scan tree trunks for prey are probable candidates. They are more likely to find peppered moths that contrast with the bark on which they rest rather than those that match the color of the bark. Kettlewell and his colleagues eventually showed that bark-scanning birds are indeed the selective agents. (Other researchers found that black moths are also better able than light moths to cope with the physiological stress that results from smoke pollution.) The importance of birds in the selection process was demonstrated in two steps. The first step was to determine if

well-camouflaged moths are really more likely to survive than poorly camouflaged ones. To this end, moths of both color forms were released on both light and dark tree trunks and later recaptured in light traps or in traps baited with pheromone-releasing females. Moths that spent the daylight hours on bark that matched their own color were more often recaptured than those that rested on bark of the "wrong" color, which indicates that the former were more likely to survive. The next step was to find out if birds are really responsible for this difference in survival rates. Moths that had been placed on matching or contrasting tree trunks were watched from blinds. Birds were often seen to capture moths that did not match the color of the bark on which they sat, but they very seldom noticed moths that did match their background.

Most animals are camouflaged, including insects and other arthropods, fish, frogs, reptiles, some birds, and many small mammals. The dead leaf butterfly of India rests among dead leaves with its wings closed and only their lower surfaces showing. True to its name, it is colored like a dead leaf, in matching shades of beige and brown. A line that runs across the undersides of both wings suggests the midrib of a leaf. The broadly expanded, green wings of katydids look like the leaves among which they rest, and the wings of some species even sport replicas of disease spots, insect nibbling, or other common blemishes. Like peppered moths, other nocturnal moths that rest on the bark of trees in the daytime are matching shades of gray and brown—except for a few white species with black markings that sit on the trunks of white-barked birches. Like these moths, rabbits rely on the camouflaging color of their coats. Instead of running away, they freeze when another animal approaches. They bolt only if they perceive the approaching animal as a threat because it comes too close for comfort. So it is with most camouflaged insects.

When they pick a place to rest, most camouflaged animals, the insects among them, choose a background that matches their appearance. Those that do not are soon eliminated by predators. Eberhard Curio, a well-known German ethologist, discovered that the leaf-eating caterpillars of a hawk moth that is found on the Galápagos Islands are always green when they are small, but when they grow larger some turn gray or brown while others remain green. Both small and large green caterpillars sit on green leaves when they rest, but gray and brown ones rest on gray or brown twigs. Other animals behave similarly. On the island of Martinique, gray, brown, and green anoles (small lizards) forage together on the ground for insects. If they are disturbed, they scatter and seem to disappear. The green ones run to green foliage, the brown ones to withered bushes with brown leaves, and the gray ones to the gray bark of trees.

Some moths that resemble bark have on their wings dark brown or black streaks that look like the crevices in bark. Different species habitually assume different postures. Some sit sideways, with the body horizontal, and others sit head up, with the body vertical. Some hold the wings out to the side like an airplane and others hold them over the body. But, generally speaking, all of them rest so that the streaks on their wings are parallel to the crevices in the bark, thus enhancing their camouflage. Under most circumstances, camouflaged animals "blow their cover" if they move. But under some circumstances, appropriate movements enhance camouflage. In *Adaptive Coloration in Animals,* Hugh B. Cott described how some leaflike butterflies, such as the dead leaf butterfly, gently sway from side to side to simulate movement in the breeze. Major Hingston described the camouflage-enhancing movements of an immature mantis that sits on the bark of a tree with its head down. Its head and thorax are gray with patches of green and yellow that simulate lichens that grow on the bark. But its gray-green abdomen, which resembles a leaf, droops down over its head and thorax. The mantis mimics the trembling of a leaf in the breeze by swaying its abdomen, "sometimes gently, at other times vigorously, just as a leaf is made to move when touched at one time by a puff of air and at another by a distinct breeze."

In the summer of 1967, Aubrey Scarbrough had a curious experience that demonstrated the survival value of camouflage. More than a dozen abnormally colored, light blue cecropia caterpillars appeared among scores of normally colored, green caterpillars that he was raising. These are the only naturally occurring blue cecropia caterpillars that have ever been reported. He had them confined with a large number of the normally colored green caterpillars under a net on a small apple tree. Before the caterpillars could spin their cocoons, an unidentified vandal with a BB gun, probably a mischievous youngster, shot all but three of the huge, conspicuous, blue

caterpillars, but none of the green ones, right through the net. The green caterpillars probably escaped the vandal's notice because they blended in so well with the green apple leaves. Birds, rather than humans, are the usual agents of natural selection that weed out poorly camouflaged caterpillars, and like humans, birds hunt by vision and they see colors.

Some camouflaged insects, including certain moths, grasshoppers, and tropical mantids—and even other animals such as frogs and lizards—fall back on flash colors to startle and confuse insect-eating birds or other predators that see through their camouflage. Nocturnal underwing moths, for example, rest motionless on the bark of a tree during daylight, their superlatively camouflaged front wings covering their brightly colored hind wings. People, and probably birds too, usually do not notice them. But when an underwing moth is noticed and disturbed by a person, bird, or other predator, it raises its inconspicuous and somberly colored front wings to reveal startingly conspicuous hind wings that are boldly striped with black and red, yellow, or white—the color is different in different species of these moths.

This display of flash colors probably serves the moth in two ways. First, the bright colors will surely startle the predator, which will most likely be a bird, possibly even frighten it off. The surprising display will at least give the bird pause, allowing the moth precious seconds in which to escape. Second, if a bird pursues the fleeing moth, it may be deceived into not noticing the moth after it lands on the bark of another tree. The pursuing bird is focused on the bright and strikingly visible flash colors of the hind wings of the flying moth. But the flash colors instantaneously disappear when the moth lands and again covers its hind wings with its front wings. The pursuing bird, which has a "search image" for a brightly colored animal, will probably not notice the now inconspicuous moth.

In some insects, startle responses are made downright frightening by large, eyelike spots on their wings. Our North American io moth has inconspicuous front wings that, when the moth is at rest, cover hind wings that are conspicuously banded and have at their center a large and realistic eyespot complete with a highlight on the pupil. If an io moth is disturbed, it suddenly lifts its front wings to reveal the staring eyespots. The moth

An underwing moth responding to a threat from a bird or some other creature by lifting its superbly camouflaged front wings to reveal its startlingly patterned hind wings.

seems to vanish and be replaced by the face of a glowering owl, a threatening apparition that could well frighten away an attacking bird.

Convincing evidence that startle displays and big eyespots really can frighten birds was discovered by A. D. Blest when he was a graduate student working on his doctorate at Oxford University. He did laboratory experiments which showed that the sudden revelation of the large eyespots on the hind wings of the peacock butterfly are responsible for evoking escape responses in birds. Attacking yellow buntings were often startled by butterflies with intact eyespots. But if their eyespots were rubbed off, these butterflies seldom evoked a startle response. Blest used a different experimental technique with captive chaffinches, great tits, yellow buntings, and reed buntings. He placed a mealworm, a beetle larva that is a delicious morsel for any of these birds, on a horizontal screen of frosted glass which he could leave blank or onto which he could project various images from below. When a bird was about to pick up a mealworm from the blank screen, he projected an image on either side of the insect. The sudden

appearance of single circles or concentric circles, which have the shape of an eye, startled the birds, but a far greater startle effect was elicited when two realistically eyelike patterns that looked remarkably like the eyes of a cross-eyed owl were projected onto the screen. The birds were little affected by the appearance of images, crosses or pairs of parallel lines, that did not resemble eyes.

While birds perceive large eyespots as a threat from an insect-eating vertebrate, they use very small eyespots as targets. When a bird pecks at an insect, it usually aims for the head end rather than the tail end. That way the insect is less likely to escape. Looking for the eyes is generally the best way to tell the head end from the tail end. Blest did a simple but convincing experiment to determine the effect of small eyespots on hungry birds. When he presented yellow buntings with mealworms that had small eyespots painted on their tail ends, the buntings usually directed their pecks at the tail end. But birds that received unpainted mealworms directed most of their pecks at the true head end.

As is to be expected, there are indeed deceptive insects that have inconspicuous true heads but have conspicuous "false heads" on their tail ends. A lanternfly from Thailand, illustrated by Wolfgang Wickler in *Mimicry in Plants and Animals,* is a noteworthy example. This insect, actually a planthopper with a sucking beak, rests with its head bowed low and largely hidden, thus obscuring its eyes and completely hiding its beak and antennae. Its tail end, elevated when the bug is at rest, bears a remarkably convincing false head. There are dummy eyespots and a large "beak" and a pair of long "antennae" that are actually appendages of the tips of the hind wings.

Many of the hairstreak butterflies, among them several North American species, have false heads on their hind wings. The tips of their hind wings bear small eyespots and long thin tails that look like antennae. Eberhard Curio made some close observations of the behavior of an Ecuadorian hairstreak. After it lands from a flight, this hairstreak swiftly flips around so that its tail end is where its head was an instant earlier. Then, holding its real antennae motionless, it moves its hind wings slightly, causing the false antennae to wiggle.

A plausible hypothesis has been advanced to explain why so many insects have false heads on their tail ends. The argument is that the false head

Two views of a tropical hairstreak butterfly showing the false head at the end of its wings

deflects the peck of a bird away from the more vulnerable true head. A bird that is deceived into pecking at a false head makes only a superficial wound and comes away with nothing more than an expendable bit of wing rather than with the whole insect. Finally, when the frightened insect makes its escape, it further confuses the bird by jumping or flying off in the direction opposite to what the bird expects. As Wolfgang Wickler wrote, "Two heads are better than one."

Most animals, including many insects, aggressively defend themselves against predators when camouflage or some other early defense fails. Bom-

bardier beetles and skunks discharge sprays of noxious chemicals as their last line of defense. Female wasps and bees, as well as some snakes, sting or strike to inject venom. As a threatened owl slashes with its talons and as a heron stabs with its sharp bill to ward off attacking predators, a beetle may use its jaws to bite, and a robber fly or a true bug may stab an attacking bird with its piercing beak. Grasshoppers and cockroaches may dissuade an attacking predator by kicking out with powerful hind legs armed with sharp spines.

Honey bees are well known for their painful stings and, especially in the case of the African race, for their belligerent defense of the colony, sometimes against innocent intruders but more often against nest robbers such as skunks, bears, birds, or, as you read in the previous chapter, honeybadgers or humans led by honeyguides. Male honey bees, the drones, are defenseless, but all females, both the workers and the queens, are armed with stingers served by venom glands and a storage sac for venom. The queens do not use their virtually barbless stingers to defend the colony. They sting only other queens, most often surplus queens that have not yet emerged from their natal cells. But only rarely do queens fight it out to determine which one will head the colony. The worker's stingers are barbed and are used mainly in colony defense. If the nest is disturbed by a large animal, thousands of workers swarm out to drive it away, and the intruding animal may be stung hundreds or even thousands of times. The barbed stingers of the workers remain embedded in the intruder's flesh, and when a worker tries to free herself, the stinger and venom sac tear loose from her body. She eventually dies, but her severed stinging apparatus will for some time continue to pump venom into the intruder's flesh. Such mass attacks are often fatal to humans, and probably to birds and other animals as well.

Although all worker bees have stingers and may come to the defense of the nest, it was recently discovered that certain workers are much quicker to respond to a threat than are others. Michael D. Breed, Gene E. Robinson, and Robert E. Page, Jr., reported in 1990 that the first bees to attack an intruder apparently constitute a previously unrecognized soldier caste—a rapid-deployment force always ready to defend the nest. These soldiers are genetically different from workers that forage for pollen and nectar, and their wings are also less worn, indicating that they fly less often than do foragers.

Most insects and other animals that sting, bite, or are otherwise noxious present warning displays that may ward off an insectivorous bird or some other predator before they have to resort to their defensive weapons. The warnings are given in self-interest, not out of a kindly intention to spare the attacking animal. It is, after all, risky to let things go so far that the defense of last resort must be used. Before striking, cottonmouths give warning by gaping to reveal the conspicuous white linings of their mouths. Rattlesnakes usually sound a warning buzz. Wasps and bees give ample warning. They are conspicuously colored, make loud buzzing noises, and zoom menacingly through the air.

Skunks present a warning display whose meaning is unmistakable. Their striking pattern of black and white is a glaringly obvious exception among the small mammals, most of which are inconspicuously camouflaged in brown or gray. Their appearance is probably warning enough for most predators. When the striped skunk, common in southern Canada and most of the United States, feels threatened, it fluffs out and raises its tail in the air as straight as an exclamation point. If the threat continues, the skunk turns its backside to the intruder, and, tail still raised, drums out a rapid tattoo on the ground with its front feet. If the intruder does not back off, the skunk sprays its chemical weapon, which not only is smelly but also irritates the eyes.

The defensive chemical spray of the bombardier beetle, although on a much smaller scale, is probably more potent, drop for drop, than is the skunk's noxious spray. This beetle's armament, well described by Thomas Eisner of Cornell University, is truly impressive. The bombardier, potential prey for birds and other small animals, is warningly colored with bright blue and orange. At the tip of its abdomen is a specialized organ that prepares and discharges the chemical defense. This organ has two reservoirs, one that contains hydrogen peroxide and another that contains highly deterrent chemicals called hydroquinones. If a bird or some other threatening predator ignores the bombardier beetle's warning coloration, the beetle raises its hind end and aims it at the predator. At the same time, the contents of these two reservoirs flow into a third chamber where, in an explosive chemical reaction, they are transformed into boiling hot defensive chemicals, called benzoquinones, which the beetle sprays toward the intruding predator. Any bird or other animal that tries to eat one of these

beetles is sprayed in the mouth. We don't know how birds respond, but toads show obvious signs of distress, gaping wide and rubbing the tongue against the ground. In the meantime, the beetle makes its escape.

The slow-moving and flightless lubber grasshopper of the southeastern United States is as noxious as the bombardier beetle and has an even more spectacular warning display. Most grasshoppers are camouflaged when seen against their usual backgrounds, but the large, heavy-bodied, yellow and black lubber is, as Murray Blum told me and as he and several colleagues wrote, "flagrantly conspicuous" and obviously warningly colored even in repose. At the sight of an intruder, it intensifies its warning display. It rises on its legs and partly raises its short but bright crimson front wings. If it is touched by a pencil point, or presumably by the beak of a bird, it raises its front wings the rest of the way and explosively discharges a noxious substance that has a persistent odor similar to the odor of creosote or carbolic acid. Blum, professor emeritus at the University of Georgia, observed that reptiles, birds, and some mammals are obviously frightened by the lubber's display and quickly back off. He also found that birds that ignore the warning and eat a lubber become ill, vomit, and will thereafter refuse even to touch one of these insects.

There are, however, birds of one exceptional species that can circumvent the lubber's formidable defense. Loggerhead shrikes, as Reuben Yosef discovered, impale these grasshoppers on a thorn or barbed wire—as they customarily do with their other prey—and eat them only after they have hung there for a day or two and their yellow markings have turned brown. Even then the shrikes consume only the head and the abdomen, discarding the thorax, which is the seat of the lubber's defense. Yosef and Douglas Whitman hypothesize that shrikes can eat lubbers because some of the noxious chemical degrades after the grasshoppers die and because the rest of it is discarded with the thorax. Eating lubber grasshoppers is very likely to be a learned rather than an instinctive behavior; experienced birds do it but inexperienced birds don't. This was demonstrated by an experiment in which Yosef and Whitman offered living lubbers to caged young and old loggerhead shrikes. Ten young, and probably inexperienced, shrikes all attacked the lubbers but were effectively repelled. They quickly dropped the lubbers and "gagged, stuck out their tongues, drooled, squawked, and shook their heads." By contrast, four of ten adult and presumably experi-

A toad gagging after trying to eat a bombardier beetle, which is making its escape

enced shrikes that had been trapped in the field did manage to handle lubbers, impaling them on a strand of barbed wire in the cage and eating only their heads and abdomens after the grasshoppers had sufficiently "ripened."

Some insects contain poisons that do not affect an insectivorous bird until after it has swallowed one of them. Although Jonah did survive his encounter with the "great fish," being swallowed is almost always fatal. How does an animal benefit by containing a poison that is effective only after the animal has been killed and eaten? This seeming paradox is resolved once we realize that birds, which are made ill by the poison but are rarely killed by it, soon learn to recognize the warning signals of poisonous insects and reject them on sight. Eating a poisonous insect is an educational experience for predators, and thus the death of the insect victim serves to protect other members of its own species. If these other members of the species are relatives of the victim—perhaps its own offspring, siblings, or first cousins—their genetic makeup will be similar to the victim's, and many of the victim's genes will survive in them. In almost all organisms, including most insects other than bees or wasps, half of a victim's genes will sur-

vive in any one of its offspring, half in a sister or brother, a quarter in a nephew or niece, and an eighth in a first cousin. Thus by sacrificing itself, a victim enhances its own inclusive fitness by making possible the survival of its genes in its relatives. The meaning of inclusive fitness becomes apparent if we view an organism as the genes' way of surviving by making copies of themselves—something like Samuel Butler's not altogether improbable aphorism that "a hen is just an egg's way of making another egg." In short, an organism's fitness is measured by the survival of its genes.

Some insects synthesize poisons within their own bodies. Others sequester poisons that are contained in plants that they eat. Monarch butterflies contain toxic substances that they acquire from milkweed plants they eat when they are caterpillars. And these butterflies advertise their toxicity by means of their blatant black and orange warning colors. The toxins from milkweeds, called cardenolides, are related to digitalis, which is used to treat heart disease, but these toxins can be fatal to birds in fairly low doses. Fortunately for an inexperienced bird that eats a monarch, milkweed cardenolides induce vomiting at a somewhat lower dose than the lethal dose and are thus usually eliminated before they kill the bird.

As Lincoln Brower demonstrated, these poisons benefit monarchs. Brower raised some monarch caterpillars on species of milkweed that do not contain cardenolides and others on milkweeds that do contain them. Captive blue jays, confined long enough to forget any previous experience that they may have had with a toxic monarch in nature, readily ate monarchs raised on poison-free plants. They continued to eat nontoxic monarchs as long as they were offered. Other jays readily ate monarchs that had been raised on toxin-containing plants, but soon thereafter they showed obvious signs of distress. They erected their crests, fluffed out their feathers, and vomited. After that one bad experience, they refused to eat either toxic or nontoxic monarchs. Some of them retched when they so much as saw a monarch.

Poisonous insects are seldom deadly to birds or other predators. An insect's inclusive fitness is not enhanced if it kills the bird that eats it. Indeed, the victim benefits if the predator survives after devouring the victim and becoming aware of its poisonous properties. Most birds, like the majority of predators, are territorial. They live in a small and delimited area from which they exclude members of their own species other than their mates. If the poisonous insect, the bird's victim, belongs to a species that does not wander widely, its surviving relatives will probably live within, or at least pass through, the territory of this bird, which will leave them alone because it has been taught not to attack them by the sacrifice of their relative. But if the territory-holding bird is killed, it will almost immediately be replaced by a wandering member of its own species, a floater that is still searching for a territory of its own. When an experienced predator that has learned not to attack the insects in question is replaced by a floater that happens to be inexperienced, the relatives of the insect victim do not benefit from its sacrifice. Thus protective systems based on benign poisons are far more likely to evolve and persist than systems based on deadly poisons.

A few poisonous or venomous insects are inconspicuous, at least to the human eye, but most of them are highly conspicuous, warningly colored as are skunks, lubber grasshoppers, and monarch butterflies. Their warning colors, sometimes just black but more often combinations of black with white, yellow, orange, or red, are a striking contrast to the background and are often accompanied by such attention-demanding behaviors as twitching movements, wing wagging, or the production of loud sounds. The strikingly conspicuous appearance of bees, wasps, and other brightly colored poisonous or venomous insects is easily remembered by predators that suffer after eating them. At least in some cases, only one such experience is enough to teach the predator not to attack the insects in question. Thereafter, the appearance of these toxic insects serves as an easily understood and unmistakable warning that is visible from a distance. The warning colors of honey bees and social wasps, alternating bands of orange and black, obviously do not deter large animals such as skunks and bears that may attack their nests, but they probably do warn off birds that may attack lone bees foraging away from the nest.

Before 1992, no bird was known to be protected by a poisonous chemical defense. In that year, John Dumbacher and his colleagues reported in *Science* that three species of pitohuis, forest-dwelling birds that occur only on New Guinea, contain a poison that was previously known only from the poison-dart frogs that South American Indians use to poison blowgun darts. In the birds, the concentration of this toxin is highest in the skin and feathers, and it caused numbness, burning, and sneezing when it contacted the mouths or nasal tissue of collectors who handled hooded pitohuis, the

species with the highest concentration of the toxin. Native New Guineans consider this species to be a "rubbish bird" that is not fit to eat unless it is skinned and specially prepared.

As can be seen in an illustration in *The Birds of New Guinea,* by Bruce Beehler, Thane Pratt, and Dale Zimmerman, two of the seven species of pitohui are obviously warningly colored. Hooded pitohuis, with the highest concentration of the toxin, are strikingly patterned with orange and black; several of the many races of the variable pitohui, which has the next highest concentration of the toxin, are almost identical in appearance to hooded pitohuis; but the rusty pitohui, perhaps in keeping with its very low concentration of the toxin, is a nondescript brown bird.

The evolution of venomous or otherwise noxious animals that emit warning signals paved the way for the subsequent evolution of harmless animals that mimic them, innocuous animals that bluff their way past predators by imitating the warning signals of some truly noxious animal. Henry W. Bates, a British naturalist, gave the first account of this sort of mimicry, based on his observations of certain butterflies in the Amazon Valley of South America. It was published in the *Transactions of the Linnaean Society of London* in 1862, only three years after Charles Darwin expounded the theory of evolution in his *Origin of Species.* The debate between the proponents of evolution and its opponents was raging. Some people opposed Darwin's theory on religious grounds, maintaining that life arose not through gradual evolution but by special creation. Some of them even maintained that all living things were created in literally six days in the year 4004 B.C.E., a date that had been calculated in the seventeenth century from biblical genealogies by Bishop James Ussher of the Anglican Church of Ireland. Bishop Ussher reckoned that the date and hour of the creation "fell upon the entrance of the night preceding the twenty third day of October." The debates were sometimes acrimonious. During one, Samuel Wilberforce, an Anglican prelate, asked Thomas H. Huxley, Darwin's foremost defender, if he was descended from an ape on his mother's side or on his father's side. In this contentious atmosphere, the Darwinists welcomed Bates's account of mimicry as an obvious and dramatic example of evolution.

Batesian mimicry, so named in honor of its discoverer, is most commonly seen in insects, but it also occurs in vertebrates. The burrowing owl's acoustic mimicry of rattlesnakes is a dramatic example. Both of these animals live in abandoned burrows in prairie dog colonies. When the owls are in a burrow, where they cannot be seen, they hiss at intruding predators, such as black-footed ferrets, making a sound similar to the buzz of a rattlesnake. But when they are above ground and it is obvious that they are not snakes, they scream and chatter at threatening predators as do other birds.

It is even possible that the coral snake, whose bite can be fatal, is a mimic rather than a model that other snakes imitate. Not only completely harmless snakes, but also certain snakes that are venomous but not lethal, are similar or nearly identical in appearance to coral snakes. Some biologists hypothesize that these moderately venomous snakes, which sicken but do not kill animals they bite, are actually the models for both the deadly coral snakes and the harmless snakes, such as milk snakes and scarlet king snakes, that resemble them. As would be expected if they really are mimics, coral snakes are reluctant to bite in defense, as is well known to snake collectors. Plate 52 of John James Audubon's elephant folio shows two chuck-will's-widows on a branch with a coral snake. Did Audubon, in ignorance of the venomous properties of this snake, handle his potentially deadly model without being bitten?

There are hundreds—most likely thousands—of examples of Batesian mimicry among insects, probably aimed mostly at birds. In the Philippines, some harmless cockroaches are dead ringers for inedible, red and black ladybird beetles; others are difficult to distinguish from toxic, yellow and black leaf beetles unless you look closely. In one species of innocuous African swallowtail butterfly, the females come in several different and widely dissimilar color forms that resemble various toxic butterflies. According to Wolfgang Wickler, a harmless tropical cricket is a convincing mimic of an exceptionally noxious tropical relative of the North American bombardier beetle. The harmless dronefly resembles a honey bee; a hawk moth mimics bumble bees; and some small, day-flying moths are mimics of wasps. Like their models, moths that mimic wasps or bees have transparent wings. They begin life with scaled wings, as do all moths and butterflies, but soon after escaping from their pupal skin they shed the scales that cover their wings and make them opaque.

*Spilomyia hamifera,* a large North American hover fly, is an unnervingly precise Batesian mimic of the fiercely stinging yellowjacket wasps. So convincing is the resemblance that *Spilomyia* sometimes deceives entomology students who should know better. I have found this fly pinned among the wasps in their insect collections. But the fly and the wasps are very different, unrelated insects that belong to different orders. *Spilomyia* has one pair of wings; the wasps have two pairs. The wasps, which are social, raise their grubs on insect prey in communal underground nests. If their nest is disturbed, the stinging wasps swarm out and mercilessly attack the intruder. In contrast to the wasps, the fly is solitary and cannot sting. As a maggot it lives in and feeds on the debris in wet cavities in trees, and as an adult it visits flowers to obtain nectar, as do the wasps.

Not only does *Spilomyia* have a convincingly wasplike color pattern, but, as I pointed out in *Evolutionary Biology,* it also mimics several other salient characteristics of wasps. Its waist is somewhat narrowed, as is the waist of a wasp. Like most other flies, it has short, bulbous antennae that are barely visible to the naked eye, but it mimics the long, black, threadlike, and highly mobile antennae of the wasps by waving its black, anterior legs in front of its head. When the wasps sit on flowers, they fold their wings and hold them out to the side. Their wings are only lightly tinted, but when they are folded lengthwise in several layers, they look like dark brown bands. The fly, too, holds its wings out to the side, but it cannot fold them. The appearance of a wasp's folded wing is mimicked by a band of dark brown pigment that runs the length of the leading edge of the fly's otherwise transparent wing. When the wasps are on flowers, they advertise their presence by rocking from side to side. The flies do not rock their bodies, but they mimic this motion by wagging their wings. Finally, when the fly is menaced by being grasped in the fingers of a human—or presumably the beak of a bird—it makes a loud squawk that is almost identical in acoustical properties to the sound made by a wasp that has been similarly menaced.

In an earlier day some biologists made the unwarranted assumption that Batesian mimics and their models must occur together in the same geographic area and at the same time. No one knows if mimics and their models can be separated geographically, but it is theoretically possible; a harmless northern insect, for example, could mimic a noxious tropical insect if both are exposed to the same birds that migrate between the north and the tropics. There is, however, indisputable evidence that mimics do not necessarily coincide with their models in time. In studies in three ecologically different and geographically widely separated habitats, I found, with the help of several colleagues, that *Spilomyia* and several other hover flies that mimic wasps or bumble bees live out their lives in early spring, when the blossoms upon which they depend for nectar are abundant. But at that time their models are almost absent. The models will not be abundant until mid to late summer, weeks after the mimics have died and are represented only by their larval offspring. We reasoned that the mimics are, nevertheless, protected by their mimicry, because in the early spring before most young birds have left their parents to feed on their own, the mimics are exposed mainly to adult birds that remember painful experiences they had with wasps and bees in a previous summer. David L. Evans, formerly one of my graduate students, compared the responses of adult and recently fledged—and thus inexperienced—birds to another hover fly, a species of *Mallota* that is a convincing mimic of bumble bees. He trapped newly returned, adult, migrant red-winged blackbirds and common grackles in spring when bumble bees were still scarce. When offered hover fly mimics of bumble bees, these adult birds rejected them on sight, even though the mimics do not sting and are perfectly edible. At about the same time, Evans took recently hatched red-winged blackbirds and common grackles from their nests and raised these nestlings in his home, getting up at all hours of the night to feed them and to make sure that they were warm. He tested the hand-raised birds when they had grown to adulthood. All of these young birds, which had never before seen a bumble bee or a mimic of a bumble bee, ate all of the bumble bee–mimicking hover flies that he offered to them—until after they had been stung by a real bumble bee. From then on they rejected bumble bees or mimics of bumble bees on sight. I do not doubt that the newly returned migrant birds rejected bumble bee mimics because they remembered painful experiences with real bumble bees during the previous summer, and that the hand-raised birds were willing to eat them because they had had no previous experience with real bumble bees.

The large and brightly colored butterflies known as pipevine swallowtails are found in southernmost Ontario and much of the United States except for the northern tier of states. (The most southern part of Ontario is south of most of the northern tier of states.) These swallowtails are extremely noxious to birds because they store poisons that are contained in the plants

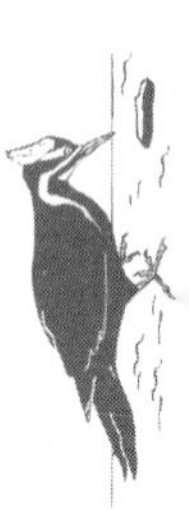

they eat when they are caterpillars. Their only food plants are members of the birthwort family, which are noted for their poisonous properties. Pipevine swallowtails feed on native birthworts, such as the Virginia snakeroot in the Midwest, but they also eat cultivated birthworts, such as the ornamental vine known as Dutchman's pipe. The striking coloration of pipevine swallowtails is a highly visible and unmistakable warning to birds and other predators. The upper surfaces of their wings are largely black and have a flashing, brilliant blue sheen that is very conspicuous when light strikes it at certain angles. The lower surfaces are black and iridescent blue, and the hind wings are marked with large orange spots.

Several familiar North American butterflies that have no protection against predators mimic the pipevine swallowtail's appearance more or less faithfully: both sexes of the spicebush swallowtail and the red-spotted purple, the female black swallowtail, the female diana fritillary, and the black color phase of the female tiger swallowtail. The day-flying males of the promethea moth are also mimics of the pipevine swallowtail, but only on the upper surfaces of their wings.

Why is it that only the females of some of these butterflies mimic the toxic pipevine swallowtail? Why aren't the males of these species protected against birds and other predators by a mimetic resemblance to this toxic butterfly? No one knows, but there are two reasonable hypotheses. One is that females gain more from mimicry than do males, because they are more exposed to insect-eating birds, perhaps because they are constantly moving from place to place as they lay their eggs. Little supporting evidence for this hypothesis has so far been found in butterflies, but the sex-limited mimicry of promethea—to be discussed below—suggests that it may well be correct. The other hypothesis is that female butterflies are less likely to accept courting males with the recently evolved mimetic pattern than males with the ancestral color pattern of their own species. There is little evidence to support this hypothesis, but the idea is that any increase in fitness that a male might gain through mimicry would be more than offset by a decrease in his ability to attract and inseminate females.

There are two color phases of the female tiger swallowtail: nonmimetic, yellow and black individuals that look like males, and black individuals that mimic pipevine swallowtails. Pipevine swallowtails are common in the southern United States, and there almost all female tiger swallowtails mimic them, except for those in a small area in southern Florida where pipevine swallowtails do not occur. North of the model's range, tiger swallowtails of both sexes are always nonmimetic. (Indeed, the northern tiger swallowtails are now considered to constitute a different species than the southern ones.) Both color phases of the tiger swallowtail occur where pipevine swallowtails are present but not abundant, as in the upper Midwest.

A person who has not closely observed tiger swallowtails under natural conditions might think that yellow and black individuals are conspicuous and assume that they are warningly colored. They certainly are conspicuous when viewed outside of their natural context, say as a dead specimen on a pin in a display case. Yellow and black bees and wasps are almost always conspicuous in nature, even at a distance. Natural selection has perfected their color pattern to be an attention-demanding warning of their ability to sting in defense. But, as contradictory as it may seem, yellow and black tiger swallowtails are quite inconspicuous under the right circumstances. The black markings on their yellow wings are arranged so as to obscure or disrupt their form. On a pinned swallowtail with its wings spread artificially wide, the black lines that traverse the front and hind wings are seen to be separate and disconnected. But when a living tiger swallowtail sits with its wings held in their natural position—the front wings partly overlapping the hind wings—the black lines on the front and hind wings come together to form continuous lines that seem to join the wings and otherwise disrupt the butterfly's form. When black and yellow tiger swallowtails sit on foliage or blossoms, they, unlike similarly colored bees and wasps, blend in quite well because of this disruptive pattern.

Jim Sternburg and I became aware of the camouflaging effect of disruptive coloration when we collected swallowtail butterflies of several species and red-spotted purples in a large field of red clover. We watched from the edge of the field and walked into the field to collect any butterflies that we saw. Not surprisingly, the mostly black-colored butterflies, including the red-spotted purple and many swallowtails, were conspicuous from a considerable distance. But the yellow and black, tiger-patterned form of the tiger swallowtail was not, although the black mimetic form of the female tiger swallowtail was easy to see. We seldom noticed the tiger-patterned butterflies from a distance, and then only when they flew. Most of the tiger-patterned butterflies we collected were spotted from close up as we walked

into the field to net a mostly black butterfly that we had seen at a distance from the edge of the field.

Why should there be two color phases of female tiger swallowtails, a mimetic one and a nonmimetic one? The mimetic color phase does not, of course, offer an advantage where there are no models to educate the predators, and where models are absent the female tiger swallowtails revert to the nonmimetic phase. Why bother to revert? There is probably a good reason. Perhaps there is some disadvantage to being mimetic. But what that disadvantage might be has yet to be discovered.

An intuitively pleasing hypothesis is that males prefer to mate with females that have the ancestral yellow and black pattern. If this is so, the advantage of being mimetic will outweigh the disadvantage of being less attractive to males only where the model is abundant. Where the model is absent, no advantage is gained by being mimetic, and natural selection eliminates the nonmimetic color phase. One researcher tried to support this hypothesis by collecting and examining both mimetic and nonmimetic females in an area where both occur. When he dissected them, he found more sperm packets in nonmimetic females, indicating that they had been more frequently inseminated than mimetic females. But the difference is small and his data are scanty and not statistically significant. Since then other researchers have found that females of the two color phases are inseminated about equally often.

My own hypothesis is that both color phases persist because each one is protective, but only under the right circumstances. In an area where the pipevine swallowtail models are sufficiently abundant, the advantage of being black and mimetic outweighs the presumably lesser advantage of having the camouflaging tiger pattern. But where models are absent, being mimetic offers no advantage and tiger swallowtails fall back on the camouflaging pattern.

The sex-limited mimicry of the palatable promethea moth, the only mimetic species among the giant silk moths of North America, is reversed. Promethea males mimic the pipevine swallowtail—at least on the upper surfaces of the wings, which are mostly black—but the light orange-brown and creamy-white females are obviously nonmimetic. In these moths, unlike the butterflies, the males, the mimetic sex, are far more exposed to insect-eating birds than are the females and thus stand to gain more from being mimetic than would females. The males of the other giant silk moths are generally not exposed to diurnal insectivorous birds, because they fly only at night. But promethea males, which are day fliers, are very much exposed to these birds. Female promethea moths, like the females of the other giant silk moths of North America, fly only when they lay their eggs at night. But unlike the other female giant silk moths, they emit their sex-attractant pheromone and mate only during the afternoon. So while the female remains safely hidden in foliage, the male must fly to her in broad daylight, thereby exposing himself to hungry, diurnal, insectivorous birds.

Seen close up, as when held in the hand, a male promethea bears only a moderately convincing resemblance to a pipevine swallowtail—and then only on the upper surface of its wings. Why doesn't the lower surface of its wings mimic the lower surface of the swallowtail's wings? The answer is probably that it is not necessary. The lower surface of a swallowtail's wings is exposed to view mainly as it sits with its wings held together up over its back when it takes nectar from a flower. But promethea males never sit on flowers, because, like all the other giant silk moths, they have vestigial mouthparts and do not feed. They live only briefly and survive on the fat stored in their bodies. Even though they are mimetic only on the upper surfaces of their wings and even though they are not highly convincing mimics when held in the hand, promethea males are all but indistinguishable from black-colored swallowtails when they are flying—and during the day they are always flying unless they are hidden in foliage. Flying promethea males have fooled me several times. One afternoon as I sat in my parked car waiting for my wife to come out of the house, I saw what I took to be a swallowtail flying across my front lawn. I did not realize that it was actually a promethea male until it tried to beat its way through a screen that blocked its access to several pheromone-releasing female prometheas that were in a cage on my porch.

Jim Sternburg, Michael R. Jeffords, and I took advantage of the promethea male's butterfly-like behavior to analyze and test the theory of Batesian mimicry under natural conditions. Many experiments with captive birds in the laboratory have demonstrated the effectiveness of Batesian

mimicry. But there had been only one previous attempt, only partly successful, to test the validity of the Batesian mimicry theory under natural conditions in the field. Promethea males were suitable subjects for our experiment, because they fly like butterflies, look like butterflies, and, like butterflies, are active during the day. But unlike butterflies, which must be recaptured individually with hand nets, promethea males can be recaptured in large numbers and with minimal effort in traps baited with pheromone-releasing females.

Jeffords painted the largely black upper wing surfaces of male prometheas in three different ways. The black wings of some were painted only with black, a control for the effect of painting the wings that did not change their appearance or alter their resemblance to the toxic pipevine swallowtail. Others, painted with yellow stripes, were caricatures of the yellow and black pattern of the nonmimetic phase of the palatable tiger swallowtail. Those in the third group were painted with orange bars to resemble the toxic monarch. Equal numbers of each group of these differently painted moths were released, in a largely wooded natural preserve in central Illinois, in the center of a mile-wide circle of seven traps baited with virgin, pheromone-releasing, female promethea moths. Many of the painted promethea males were recaptured in the traps. The recaptured males had run the gauntlet of insectivorous birds as they flew at least half a mile from the center of the circle to one of the traps, passing through a woodland of mixed hardwood trees and over a small area of grassy prairie.

The results support the hypothesis that mimetic butterflies are protected by their resemblance to warningly colored toxic butterflies. Moths painted with orange to resemble toxic monarchs or with black to retain their resemblance to poisonous pipevine swallowtails were more likely to survive and be recaptured than were moths painted to resemble the edible yellow and black tiger swallowtail. Quite a few of the recaptured yellow-painted moths bore wing damage that had, without a doubt, been inflicted as they escaped from an attacking bird. Large pieces had been torn from the wings of some, and the triangular imprint of a bird's beak was clearly visible on the wings of a few others. Bird-inflicted injuries seldom occurred on the wings of moths that had been painted to resemble either toxic pipevine swallowtails or monarchs. Birds, probably the only predators to attack the flying moths, were apparently able to distinguish between painted moths that looked like toxic butterflies and painted moths that looked like edible butterflies.

The monarch and the viceroy butterflies are only distantly related, but they are so similar in appearance that they are difficult to tell apart unless you take a very close look. (Interestingly enough, the viceroy deviates greatly from the appearance of its near relatives. None of them resemble monarchs, and most of them have blackish wings crossed by broad, white disruptive bands.) The viceroy was long thought to be a classical example of a Batesian mimic, palatable and protected only by its resemblance to the monarch. But this is not really the case. As first reported by Jan Van Zandt Brower in 1959, the viceroy is sometimes rejected by birds, although not as often as the toxic monarch. It does not, therefore, strictly fit the definition of a Batesian mimic, an edible animal that is bluffing. It is, rather, an example of Müllerian mimicry, named for Fritz Müller, who discovered a previously unknown type of mimicry that is fundamentally different from Batesian mimicry: a close resemblance between two animals, both of which are toxic and warningly colored. In an 1879 publication, he argued that two or more toxic species will tend to evolve similar warning signals and thus come to resemble each other. Müller contended, quite reasonably, that even in the most toxic and flagrantly conspicuous species, some individuals will always be killed in the process of educating and reeducating predators. Consequently, if two species have similar warning signals, there will be an economy for both, because the inevitable mortality will be shared among a larger number of individuals. Müllerian mimicry is common and widespread. The stinging wasps are a familiar and ubiquitous example. Their species number in the thousands, but almost all share the same conspicuous "advertising logo." Their bodies are marked with alternating rings of yellow and black, and they make similar buzzing squawks when threatened.

There is a third type of mimicry known as aggressive mimicry. Camouflage protects insects and other tiny creatures against predators. But camouflage also serves insect eaters and other small carnivores in two ways: it makes them less visible to their own prey, and it tends to hide them from birds, lizards, and other predators that would gobble them up on sight.

North American mantises are well camouflaged. Green ones ambush prey from a resting place among leaves; brown and gray ones generally lie in wait on similarly colored bark. But on the Malay Peninsula there is an even more deceptively camouflaged mantis. It lies in wait for its prey among the

bright pink blossoms of a plant known as the *sendudok* in the Malay language. Almost its entire body is bright pink, and its middle and hind legs are adorned by broad, thin flanges that resemble the pink petals of a *sendudok* blossom. Its appearance is so deceptive that nectar-seeking flies sometimes land on its body rather than on the surrounding flowers. In a 1900 report of a Cambridge University expedition to the Malay Peninsula, Nelson Annandale described this mantis, explained its behavior, and related some of the Malay myths that surround it. The natives of Aring, a village in southern Siam (now Thailand) believed that the *kanchong,* the Malay name for this mantis, is

> but a flower which has become alive. Its origin is from the flowers. The blossoms of the "*Sendudok*" give birth to it in the same way as the leaves of the "*Nanka*", or Jack Fruit tree . . . give birth to . . . a large prickly [stick insect] of great rarity, which rich men keep alive in cages in order to secure its [bright red] eggs, which they set in rings like jewels, and consider to be a most powerful charm against evil spirits of all kinds.

There are other examples of such treachery among animals other than insects and even among plants. A giant alligator snapping turtle lying with all but its head hidden in the mud at the bottom of the Mississippi River, jaws gaping wide and wormlike red tongue wriggling, is surely using aggressive mimicry to attract a fish that will become its next meal. Some North American freshwater clams use an even more bizarre form of aggressive mimicry. As larvae they are tiny parasites on the gills or fins of fish, but eventually they drop off to burrow into the bottom sediment and grow to be big clams. Fully grown females discharge thousands of these parasitic larvae. Those larvae that live in still water lie on the bottom until a fish inadvertently sucks them up as it gulps a bit of food from the bottom. But species which live in fast currents that will wash away their larvae have adopted a different strategy. They use a remarkable fishlike lure to attract large fish that are suitable as hosts for their larval offspring. The lure, a fleshy portion of the clam, protrudes from the gap between the two valves of the female's shell. It has a tail, fins, and two eyelike spots on its head. It faces upstream and undulates like a small fish. When a large, fish-eating fish is fooled by the lure and comes close, the clam releases a puff of parasitic larvae into its face. Lucky larvae attach themselves to the fish and begin the parasitic phase of their lives.

Even plants use aggressive mimicry. Some plants produce the odor of rotting meat and trick carrion-feeding flies into pollinating them. The blossoms of the bee and wasp orchids of Europe are approximate but recognizable visual mimics of their namesakes, and give off scents that mimic the sex-attractant pheromones of wasps or bees. Male wasps or bees make futile attempts to copulate with the blossoms, in the process acquiring bundles of pollen that stick to their heads and will fertilize the next orchid blossom that deceives them.

While the sexually frustrated wasps and bees are victims of aggressive mimicry, other insects—and some spiders, too—use aggressive mimicry to capture their prey, as does the flower-mimicking mantis. If an insect is the intended victim, the falsified signal is more than likely to be chemical, an odor or a flavor, rather than aural or visual, since, as you know, many insects are deaf and few see well.

Among the most remarkable of the arthropodan aggressive mimics are the nocturnal bolas spiders. Bolas spiders prey only on the males of certain night-flying moths, which they attract from a distance. The moths approach the spider from downwind, and they can be caught in traps baited with bolas spiders that are out of sight. These observations lead to the inescapable conclusion that bolas spiders attract the male moths that are their prey by mimicking the odor of the sex-attractant pheromone of female moths. In a 1987 article in *Science,* Mark K. Stowe and his colleagues presented direct evidence that hunting bolas spiders really do synthesize and release several chemical compounds that are identical to components of the sex-attractant pheromones of the moths on which they prey.

Bolas spiders capture the moths that approach them in a way that is unique among animals—except that humans have invented similar techniques, the use of the bolas and the lasso. A hunting bolas spider suspends itself from a few strands of silk that it attaches to the surrounding vegetation. The "bolas," a separate strand of silk with a droplet of very sticky glue at its end, dangles from one of its legs. When a moth comes close enough, the spider flicks the bolas at it. If the spider's aim is good, the moth sticks to the glue and the spider reels it in.

The insect-eating crab spiders, familiar to both European and North

American naturalists, wait in ambush to seize their prey. They do not spin webs. Some species wait on blossoms to catch bees, flies, butterflies, or other insects that come for nectar or pollen, and some can even change color to match the blossoms on which they lurk. One crab spider can switch between yellow and white and another can change from pink to yellow and back again. In *The World of Spiders,* W. S. Bristowe, a British arachnologist, described a simple experiment that provides strong support for the proposition that the crab spider's camouflage is an aid in catching prey. In the center of each one of 16 yellow dandelion blossoms that he had arranged on a lawn, Bristowe placed a pebble of about the same size as a crab spider. Half the pebbles were black and conspicuous, and the other half, about the same color as a crab spider and a match for the yellow of the dandelion, were very inconspicuous. Bristowe watched these dandelion heads for half an hour—with the astonishing result that only 7 insects visited the ones with black pebbles, while 56 bees and flies visited the ones with yellow pebbles. This experiment persuaded Bristowe that crab spiders profit by matching the colors of the flower on which they sit and thereby not alerting the insects they are waiting to ambush. This experiment did not prove that camouflage protects crab spiders from animals that prey on them. But Bristowe noted that among 10,000 spiders found in the stomachs of birds by the U.S. Biological Survey, only 4 were crab spiders.

# Bugs That Eat Birds

# 5

It was a beautiful spring morning in 1984. Later that day I would drive my car onto the Texas Highway Department's ferry to cross from Port Bolivar to Galveston as I birded my way south to Brownsville and the lower Rio Grande Valley. As the ferry crossed Galveston Bay, I would see many birds, including sandwich terns and a magnificent frigate bird. But there would be no brown pelicans. The first time I made this crossing—in the summer of 1960—brown pelicans had been everywhere: fishing the waters of the bay and sitting on almost every piling at the ferry slips. But by 1984 the brown pelicans along the coasts of Texas and Louisiana had been all but exterminated by nonbiodegradable insecticides that had made their way into coastal food chains from nearby rice and cotton fields. As I was told by my friend and fellow entomologist Jimmy Olson of the Texas Agricultural and Mechanical University, the farmers now use less persistent insecticides, and consequently brown pelican populations are recovering all along the Texas and Louisiana coasts. On a 1992 visit to the Galveston area, I saw brown pelicans, but they were not yet as abundant as they once had been.

But on that morning in 1984 I birded the fabulous Bolivar Flats, haven for migrating shorebirds and mecca for birders. It was a good morning. The tide was out and flocks of shorebirds were everywhere on the exposed, sandy flats. I watched a group of American avocets wading abreast through the shallows as they scythed their long, upturned bills from side to side through the water in a cooperative hunt for small marine organisms. I saw my lifer Wilson's plovers on the dry sand behind the beach. I watched a black-necked stilt as it preened itself at the edge of a nearby flooded field. It stood on one long, red leg as it scratched its head with the other leg, which reached forward as it passed over the drooping wing. Seeing that bird scratch brought out the entomologist in me. I wondered if the scratching was the stilt's response to the presence of a pestiferous louse or mite.

There are certainly many reasons why birds scratch. One of them is to better distribute the oil that is smeared onto the head when the head is rubbed against the preen gland on the rump. But another reason—one that leads to the subject of this chapter—is to relieve an itch. Birds may be itchy for many reasons, but often the cause is the activities of bugs that eat birds—parasitic mites or lice that live on their bodies or mosquitoes or other insects that visit them briefly to suck their blood. The skin of a bird is different from ours; it has fewer nerve endings, a fact that led one of my ornithological colleagues to question whether or not birds can perceive an itch. But birds do scratch if they are lousy or infested with mites. Some sensation must alert them to the presence of these pests. This sensation may not be the same as what we feel, but it seems reasonable to call it an "itch" since it elicits a scratch.

Lice are of two kinds: sucking lice and biting lice. Since all the sucking lice inhabit only mammals, they will be of little concern here. They use their highly specialized mouthparts—with the mandibles reduced to mere vestiges—to pierce the skin of a mammal in order to suck its blood, their only food. Most of the biting lice live on birds, although a few of them infest mammals. Their mouthparts, unlike those of the sucking lice, are basically of the chewing type, always include functional mandibles, and, in most species, are used to eat solid food.

The solid food of the biting lice includes the barbules of feathers—especially down feathers—which are cut into bite-size pieces by the mandibles. Their food also includes bits of dry, sloughed-off skin and other organic debris found in or under the plumage of a bird. Some biting lice consume blood. But with a very few exceptions, the mouthparts of biting lice that drink the blood of birds are not modified for piercing and sucking. Most of these blood feeders simply lap up the blood that oozes forth when they use their mandibles to lacerate the skin or to puncture the quills of a developing feather. But according to Theresa Clay, an authority on these insects, some biting lice, a few species that live on hummingbirds, do have their mouthparts modified for piercing and, presumably, for sucking blood. The mouthparts of these blood-eating biting lice are, however, far less specialized and fundamentally different in structure from those of the sucking lice, and un-

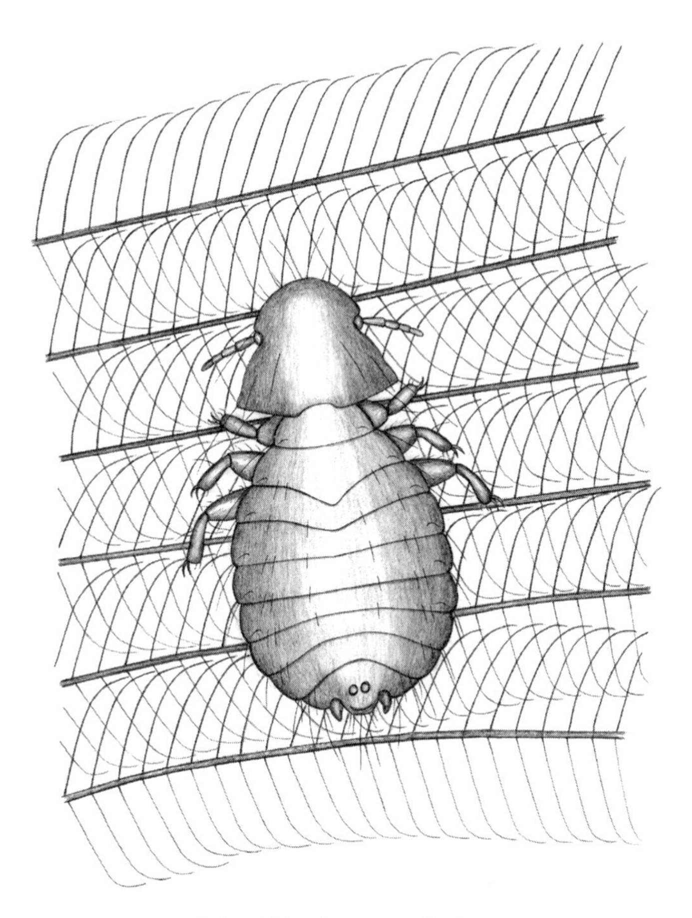

A tiny biting louse on a feather

like the sucking lice, they have mandibles. Although the mouthparts of these lice of hummingbirds are roughly similar in *function* to those of the sucking lice, the basic differences in their *structure* tell us that the two groups evolved the blood-feeding habit independently of each other.

In his authoritative book *Parasitic Insects,* R. R. Askew points out that "of all the insects, lice are the most completely committed to parasitism . . . for lice the bird or mammal host is the only environment." All life stages—egg, nymph, and adult—are spent on the host's body. A biting louse may spend its whole life on the body of the same bird, and its descendants for several generations may remain on that bird. Some of these descendants must,

however, at some time or another, transfer from the bird of their birth to some other bird—almost always another individual of the same species. If they do not, their line will die out when their host bird dies. Since all lice are wingless, the transfer can take place only when two birds are in close bodily contact, except, as I will discuss later, when the louse is transported by some other creature. Transfers may occur when the birds mate, as they care for the young, or, in some species, when individuals are closely crowded together as they roost. In rare instances, the transfer may be from a bird of one species to a bird of a different species. Lice do occasionally transfer from their host to the body of a predator—perhaps a goshawk or a falcon—as the predator eats the original host. They seldom survive on the predator. But a louse that parasitizes both falcons and pigeons appears to be an exception. It is generally thought that this louse came to live on two such distantly related hosts because of transfers from prey to predator.

As is to be expected, lice are well adapted to live and survive on the bird or other animal of their choice. Their anatomy, behavior, and life history are all geared to life on their host. Their winglessness and flattened bodies are obvious adaptations for living among feathers or fur. Since all lice undergo gradual metamorphosis and thus have virtually identical requirements as nymphs and adults, they are ideally suited to spend their entire lives, both as immatures and adults, on an animal. Mating must, of necessity, occur on the host's body. If kept in the laboratory, lice do best at a temperature that is close to the temperature at the surface of the host's body. For instance, the eggs of the chicken louse hatch after 3 to 5 days if kept in an incubator at about 99° F. (The internal body temperature of a chicken is about 107°, but the temperature at the body surface is much lower.) Slightly lower temperatures prolong the hatching period to 14 days, and nymphs soon die if the temperature is lowered by only 9° to 90°F.

The eggs of biting lice are firmly glued to the feathers of the host bird by a cementlike substance secreted by the female. Biting lice that live on the wings of birds place the eggs end to end in rows in the grooves between the barbs of the flight feathers or the underwing coverts. Here they are protected by the raised edges of the barbs from the wipes of a preening bird's bill. The eggs of lice that live on the head or neck of the bird cannot be reached by the bill—they can only be reached by the bird's claws, which are relatively inefficient at removing parasites. These head lice lay their eggs singly or in small clusters at the base of a feather. Nymphal lice feed as do their parents, and development from egg to adult is generally completed in about a month. The constant warmth of the host's body makes it possible for lice to produce generation after generation throughout the year even in cold climates.

Some kinds of biting lice can live on several unrelated species of birds, but many others are host specific and can live only on one or a few closely related species. In *Parasitic Insects,* R. R. Askew points to a good example of specificity, the restriction of the biting lice of the genus *Philopterus* to British crows and crowlike birds of the genus *Corvus, Philopterus atratus* on the rook, *P. corvi* on the raven, *P. guttatus* on the jackdaw, and *P. ocellatus* on the carrion crow and the hooded crow, two subspecies of *Corvus corone.*

The biting lice of cuckoos are another example of specificity. Of the 143 species of cuckoos in the world, about 60 percent (including all the North American species) usually build nests and incubate their own eggs, but the other 40 percent (mostly Eurasian, African, and Australian species) lay their eggs in the nests of other birds and leave them there to be cared for by the unwitting foster parents, just as do the unrelated New World cowbirds. The many species of lice that infest the numerous passerine foster parents of the common cuckoo of Eurasia never establish themselves on parasitic cuckoo nestlings. Although cuckoos fool the foster parents of their young, they do not deceive the lice that infest the foster parents. As Miriam Rothschild and Theresa Clay point out in their fascinating book *Fleas, Flukes, and Cuckoos,* the common cuckoo is inhabited in England by only three species of lice, which belong to the aptly named genera *Cuculoecus* (inhabitant of cuckoos), *Cuculicola* (dweller on cuckoos), and *Cuculiphilus* (lover of cuckoos), genera that are found on cuckoos throughout the world but that never infest other birds, not even the songbirds that are the foster parents of these cuckoos.

Much like certain North American wood warblers that manage to coexist in the same northern coniferous forest by occupying slightly different ecological niches, several species of biting lice may split up the available territory by occupying the different ecological niches offered by the body of the same bird. V. A. Dogiel, a Russian parasitologist, made this point by illustrating how several species of biting lice partition themselves on the body of the glossy ibis, a bird that occurs both in the Old World and the New World.

One species occurs only on the feathers of the head and neck; another only on the feathers of the breast, belly, and sides of the abdomen; a third only on the back; and finally, two or more species inhabit the feathers of the wings and tail.

In an article on the evolutionary origin of the biting lice, Theresa Clay showed not only that these insects are marvelously adapted for their parasitic existence on the body of a bird, but that they are precisely adapted to survive in the particular niche that they occupy. Speaking of birds and biting lice in general, she observed that "on the head and neck . . . is found a short, round-bodied type, not greatly flattened . . ., and with a large head to accommodate the enlarged mandibles and their strong supporting framework." She goes on to argue that lice that live on the head and neck can afford to be relatively thick-bodied because they are out of reach of the bird's bill. If they were on some part of the body that is accessible to the bill they would easily be picked off or crushed as the bird preens. By contrast, biting lice that live on the wings and body of a bird are sleek, greatly flattened, elongate, and agile, built so that they can hide in the grooves between the barbs of feathers, and also capable of avoiding the preening bill by rapidly slipping sideways across the feathers.

The pelicans and some of the cormorants are infested by various species of biting lice of the genus *Piagetiella,* all of which occur only on these two families of fishing birds of the order Pelicaniformes. The lice of this genus live within the throat pouches, attached to the inner wall. What they eat is not known with certainty, but the usual conjecture is that they eat blood, other body fluids, and perhaps solid debris from the walls of the pouch. They presumably avoid being swallowed or discarded with the water that enters the pouch when a fish is caught by holding on tightly.

As to the evolutionary origin of the lice, entomologists are generally agreed that they and the psocids, loosely known as the booklice and bark-lice although they are not really lice, evolved from a common ancestor that probably looked and behaved much like a modern-day psocid. This theory was advanced by the American entomologist A. S. Packard in 1887 and was recently reiterated by Stephen C. Barker, an Australian entomologist at the University of Queensland. As Barker pointed out, many psocids "live in leaf litter and detritus and feed on fungus, algae, pollen, and other organic matter." Not surprisingly, some psocids are occasionally found in birds' nests, where they presumably eat organic detritus, including the bits of dry skin and other matter that fall from the bodies of the birds that occupy or occupied the nest. The theory is that the ancestor of the lice was most likely a nest-inhabiting psocid that climbed onto the body of the resident bird to gain more direct access to the bird dandruff that was its principal food. This could well have been the first step in the evolution of a true parasitic association between the birds and the ancestors of the modern lice. In 1967, Edward L. Mockford, the world authority on psocids, reported that psocids actually do sometimes climb onto the bodies of birds. Normally free-living insects of this order have been found on the bodies of birds from Korea and the Philippines: a bunting, a ground dove, a drongo, and a kingfisher.

A bird is inevitably weakened by infestations of lice or other parasites or by infection with disease-causing microbes. It would certainly benefit a mate-seeking bird to be able to choose a healthy bird, neither parasitized nor diseased, for a mate. A parasitized mate is a liability, likely to be weak and certainly a source of parasites that will probably infest its spouse and its young. In species that share parental care, the survival of the young may be jeopardized by an unhealthy or heavily parasitized mate that is too weak to carry its share of the parental burden. Even in species in which the female is solely responsible for parental care, as is the case with grouse and other gallinaceous birds, an unhealthy father may lack genes for resistance to disease or defense against parasites—genes that may be present in a healthy male and that may be passed on to his offspring.

During the summer of 1982, I was happily doing research at the University of Michigan's Biological Station—affectionately called the bug camp by local residents—on beautiful Douglas Lake in northern Michigan. It was an exciting time. The surroundings were wild and lovely; the enthusiasm of the students, faculty, and visiting researchers was contagious; and my research on mimicry was going very well. That summer William Hamilton and Marlene Zuk were working at the station. They had just formulated the new and exciting hypothesis that some of the secondary sexual characteristics of male birds, such as bright plumage, the combs and wattles of roosters, and the air sacs of grouse, function not only as "ornaments" to entice females, but also as highly visible barometers that reveal the state of

a male's health. A healthy rooster, for example, will have a bright red comb and wattles, while those of a sick or heavily parasitized rooster will be pale and washed out.

Hamilton and Zuk had found that there is a positive correlation between the brightness of a species' plumage and the proportion of the individuals of that species that are infected by blood parasites. This suggests that species that tend to be sickly are more likely to evolve bright plumage, perhaps because individuals of that species can benefit by rejecting as mates unhealthy individuals whose plumage is not up to par.

The correlation found by Hamilton and Zuk is certainly suggestive but does not prove that birds use visual cues, such as the condition of a male's secondary sexual characteristics, to reject unhealthy potential mates. More data are needed. In an observational study of Danish barn swallows—the same species that occurs in North America—Anders Møller found additional evidence for the Hamilton-Zuk hypothesis. Both male and female barn swallows have long outer tail feathers, but those of the males are even longer than those of the females. Males infested with mites and biting lice have shorter outer tail feathers than do males that are free of parasites. Females are more likely to mate with long-tailed than short-tailed males. Møller found that the choosiness of the females appears to pay off. Pairs that are parasite-free or that carry minimal parasite loads tend to produce more, larger, and healthier offspring than do heavily parasitized pairs. Møller's findings are more persuasive than those of Hamilton and Zuk, but they are still based on correlations, and correlations are not absolute proof.

But the work of Margo Spurrier of the University of Wyoming with sage grouse does provide convincing experimental evidence supporting the Hamilton-Zuk hypothesis. Sage grouse mate on leks, areas where many males compete for each visiting female as they strut and posture before her, fluffing up their feathers, fanning out their tails, and inflating two large, bare-skinned, yellow air sacs on either side of the throat. Linda Johnson and Mark Boyce, both of the University of Wyoming, had already shown that male sage grouse infested with biting lice are less likely to win copulations than are louse-free males. The experiments done by Spurrier and her colleagues showed that female sage grouse can and do assess the parasite load of a courting male by means of the condition of his air sacs, and that they reject parasitized males. The lice cause the formation of conspicuous hematomas, blood-red spots, on the air sacs of males. Spurrier and her co-researchers thought that the presence of these spots is the clue that enables females to reject lousy males. They proved their point experimentally by showing that females will reject louse-free males whose air sacs have been marked with red spots by means of a felt-tipped pen.

The bed bug *Cimex lectularius,* commonly thought of as *the* bed bug, lives in human dwellings and creeps out at night to suck the blood of the inhabitants as they sleep. During the day it lies hidden and inactive in cracks and crannies, often under loose wallpaper, behind baseboards, or in narrow crevices in wooden bedsteads. Today bed bugs are not often seen in industrialized countries, but they were all too common a few years ago. In the 1940s, when I was growing up on the wrong side of the tracks in Bridgeport, Connecticut, the first thing that you did before renting an apartment was to check for bed bugs. It was estimated that in 1939 4 million people were plagued by bed bugs in London, and in 1865 Frank Cowan reported that Messrs. Tiffin and Son of London had hanging over the door to their shop a transparency lit by a gas flame. It said: "May the Destroyers of Peace be Destroyed by us, Tiffin & Son, Bug-Destroyers to her Majesty."

The bed bug that attacks humans does occasionally attack birds, such as house sparrows and pigeons that are associated with humans. But at last count there were known to science 26 other species of bed bugs (family Cimicidae) that are exclusively or almost exclusively associated with birds, hiding in their nests during the day and coming out at night to suck their blood.

All bed bugs are wingless, and thus cannot search for their hosts over long distances. They are occasionally carried about on the bodies of their host, but they assure their continued association with their hosts by utilizing only hosts that have permanent dwellings or return to the same nest year after year. No bed bug attacks primates other than humans, presumably because humans are the only primates that inhabit permanent dwellings. Of the 89 known species of bed bugs, 61 are associated with cave-dwelling bats and 2 with humans. Most of the 26 species that attack birds are associated with swifts or swallows, but others are associated with domestic fowl, domestic pigeons, Old World flycatchers, parrots, New World vultures, barn owls, typical owls, eagles, falcons, and other species that reuse their nests.

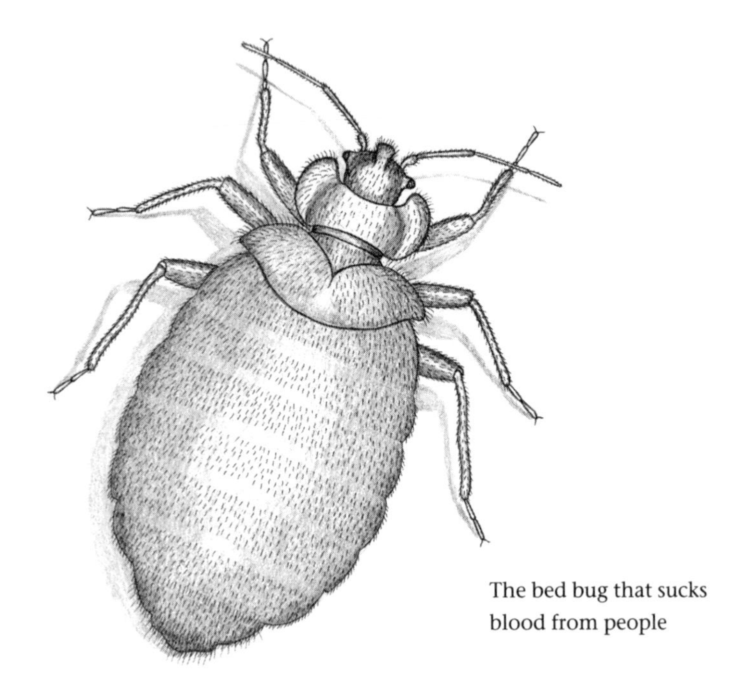

The bed bug that sucks blood from people

A species of bed bug, aptly known as the swallow bug, infests nests of the colonial cliff swallow and takes blood meals from the nestlings. Charles and Mary Brown found that swallow bugs are more numerous in large cliff swallow colonies, where the nests are densely packed, than in small colonies, where the nests are more widely dispersed. By treating some nests with an insecticide, which did not harm the swallows but did eliminate the bugs, the Browns showed that the bugs had an adverse effect on nestlings—the greater the number of bugs in a nest, the greater the adverse effect. Nestlings from treated, parasite-free nests were large and healthy, but those from untreated, heavily infested nests were small, sickly, and often died. On average, nestlings from untreated nests, which ranged from lightly to heavily infested, weighed more than 15 percent less than nestlings from nests that contained no swallow bugs because they had been treated with the insecticide.

⁎ In the 1950s, when I was an undergraduate at the University of Massachusetts in Amherst, I had the good fortune to hear Robert Frost read some of his poems. I probably heard him read "A Considerable Speck," but I can't be sure, because I have read that poem so many times that it is fixed in my mind. Perhaps I only imagine his voice reciting the lines:

> A speck that would have been beneath my sight
> On any but a paper sheet so white
> Set off across what I had written there.
> And I had idly poised my pen in air
> To stop it with a period of ink
> When something strange about it made me think
> This was no dust speck by my breathing blown,
> But unmistakably a living mite
> With inclinations it could call its own.

Frost goes on to reflect on this mite's awareness of its surroundings, recognizes that it has a mind, and concludes the poem:

> I have a mind myself and recognize
> Mind when I meet with it in any guise.
> No one can know how glad I am to find
> On any sheet the least display of mind.

As the last four lines make plain, Frost used the story of his encounter with the minute arachnid to make a critical comment—perhaps on the state of poetry or literature in its broad sense, or perhaps only on the writing abilities of his undergraduate students at nearby Amherst College. Although Frost had little or no biological interest in the mite, his poem does suggest some notable things about these creatures.

As Frost made clear, mites can be very small and almost beneath notice. A few mites, notably those known as ticks, are large enough to be readily apparent to the naked eye, but the great majority of them are minute—so inconspicuous that they often escape detection by even trained naturalists. That is probably one of the reasons why we know so little about them. Although experts estimate their true number at as many as 500,000 species, only from 20,000 to 30,000 have thus far been named. Furthermore, we know little about the natural history and behavior of these nearly microscopic arachnids. But there are some notable exceptions. We know quite a bit about mites that cause us economic losses by attacking crop

plants or domestic animals or by attacking us or transmitting diseases that affect us.

Where did Frost's mite come from? Since mites are ubiquitous, it could have come from almost anywhere. Perhaps it was blown in through the window. Perhaps it was a household mite, maybe a cheese mite or a flour mite that strayed from the kitchen, or one of the mites that live in the dust that accumulates under beds, or even a predaceous mite that feeds on these dust mites.

Mites occur in virtually every ecological niche you can imagine. They live in the soil and humus by the millions, acting as scavengers, feeding on the filamentous underground growth of fungi, or preying on other mites and tiny creatures. All kinds of plants are attacked by mites, such as the spider mites that live in greenhouses and those that feed on the foliage of orchard trees. About a third of the mites are parasitic, according to an estimate by Miriam Rothschild and Theresa Clay. They parasitize virtually all animals. Mites of the genus *Demodex* live in the pores and hair follicles of many mammals, including dogs, cats, mice, swine, and people. The late William B. Nutting, one of my professors at the University of Massachusetts and an authority on *Demodex,* estimated that about 75 percent of all people harbor these microscopic and usually harmless creatures in the pores at the base of the nose or in the hair follicles of the eyebrows. When I took one of Professor Nutting's courses, he had the students take scrapings from the base of their noses and examine them under a microscope. Sure enough, about three quarters of us found *Demodex* mites.

Other mites infest all sorts of vertebrates other than mammals, including snakes, lizards, and birds. Some mites are parasites of freshwater mussels, and in 1939 Karl Viets described a mite found in the gut of a sea urchin dredged up from the ocean depths. Insects of all kinds serve as hosts for mites: beetles, ants, flies, wasps, dragonflies, and many others. There is a tiny species of mite that lives in the tracheas of honey bees. I have seen mites clinging to the thorax of a mosquito, always with equal numbers on the two sides of the mosquito lest they disadvantage their host, and consequently themselves, by putting the mosquito off balance and thus hampering its ability to fly. But mites that live in the ears of certain moths place themselves asymmetrically, invading only one ear so as not to doom the host and themselves by completely deafening the host so that it cannot hear the ultrasonic cries of pursuing bats.

Birds are attacked by many different kinds of mites. Some of them are nest inhabitants that visit a bird from time to time in order to obtain a blood meal, others are permanent external parasites that live on the skin or among the feathers, and some are internal parasites that live within the body of a bird. Some of the external parasites feed on blood, others live within developing feathers, and some are essentially harmless, living on the outside of the feathers and eating only bird dander and other organic debris that accumulates in the plumage. Living among these parasitic mites are predaceous mites that roam through the plumage seeking the parasitic mites that are their prey.

Parasitic mites often occur on only a single species of bird or on a few closely related species. Such host specificity seems to be widespread among the mites that infest birds, but the picture is still incomplete because so many species remain undescribed and because our knowledge of the natural history of most of the described species is still fragmentary or even non-existent.

Nest-dwelling mites are associated with many kinds of birds, probably with most of the species in the world. Some of the mites that live in bird nests are not parasites that attack the bird, but are, rather, scavengers that eat only the organic debris in the nest. Other species, notably some members of the family Dermanyssidae, are parasites that take blood meals from the occupants of the nest. Among the mites of this family we see a spectrum of habits, a transition from blood-feeding nest inhabitants that visit a nest occupant only when they need a meal to permanent blood-feeding ectoparasites (external parasites) that spend their entire lives on the body of the host bird.

Relatively little is known about the many dermanyssid mites that infest birds, except for a few species that are economically important pests of poultry. The many others that parasitize wild birds doubtlessly have similar habits and are as injurious to their hosts as are poultry mites to their hosts, seriously diminishing growth and egg production and sometimes even causing the death of a bird.

When they are associated with chickens, especially in old-fashioned henhouses, chicken mites spend the daylight hours hiding under debris or in cracks and crevices. At night they come out and crawl onto birds to suck blood, their only food. Chicken mites are gray in color but appear to be red after they fill up with blood. When they are numerous, too numerous to be

accommodated by their hiding places, they appear as large patches of red around the roost or in the nest. Their eggs, only two or three dozen per female, are tucked away in the hiding places of their parents. After hatching, the mites undergo three molts as they mature to the adult stage in a period of only 7 to 10 days, their growth fueled by large blood meals that they take just before the second and third molts. Adult females continue to take blood meals from the chickens for several weeks, until they have laid their full complement of eggs.

At the other end of the spectrum of parasitism is another dermanyssid, the northern fowl mite. Like the chicken mite, this species is a blood feeder, but it generally spends its entire life on the bird and lays its eggs on the down feathers. Not only is it a significant pest of poultry, but it also attacks many wild birds, among them blackbirds, kingbirds, meadowlarks, grackles, American robins, European starlings, and sparrows.

Members of another group of mites, a motley group loosely known as feather mites, do not drink blood but, rather, eat the feathers and the horny layers of the skin as well as organic debris in the birds' plumage. Except for penguins and cassowaries, some birds of every order are known to be infested by feather mites, but at the present state of our knowledge, we can say neither how many different kinds there are nor how many species of birds harbor them. A great variety of mites have adopted this way of life, members of at least 19 different families. There are certainly many hundreds—probably many thousands of species—most of them undiscovered, and there is little doubt that acarologists (students of ticks and mites) will ultimately find that most if not almost all kinds of birds are inhabited by one of more species of feather mite. For example, when T. M. Perez and W. T. Atyeo collected feather mites from orange-fronted parakeets in western Mexico in 1984, they found that these birds were infested with 17 different species and that 15 of them were new to science.

Many species of mites are to be found on the surface of a feather, but others live within the quill. The quill-inhabiting species invade a new, developing feather and feed on its inner pulp. As the feather matures, they grow to adulthood and ultimately produce offspring. The mature females of the new generation leave the quill of the now mature feather and disperse to developing feathers on the same bird or a young bird that is being brooded in the nest or under the wing of an adult. Miriam Rothschild and

Theresa Clay postulated that quill mites "seem to know when the moult is due, for they are never found in cast feathers."

Many of the externally dwelling feather mites, pressed by competition from other species, have become so specialized that several species of them can coexist on the same bird, because, like biting lice, they have specialized to occupy the different habitats that the feathers provide. Some of them live on the down feathers, others on the contour feathers, and some even in the sometimes turbulent habitat provided by the flight feathers. Several different species may even share the same feather by occupying the different microhabitats that it offers. In an interesting 1979 article, the first to deal with feather mites of the cranes of the New World, W. T. Atyeo and R. M. Windingstad described an instance of such resource partitioning. From sandhill cranes captured with rocket nets in Wisconsin and at the Jasper-Pulaski Wildlife Refuge in Jasper County, Indiana, these scientists collected four species of feather mites, three of them new to science. All four of these mites live on the primary feathers of the wings. Two species manage to share the undersurface of the blade (vane) of the feather near its tip. One of them, a rare species, occurs only on one side of the midrib of the feather, and the other, the more common of the two, occurs on both sides of the midrib, thus occupying both the left and right sides of the blade. Another rare species is found only on the undersurface of the blade close to the base of the feather. The remaining species, a common one, occurs on both the upper and lower surfaces of the feather, on both sides of the midrib, very close to it and along most of the length of the feather except near its tip.

Mites are much less likely to be transmitters of bacterial or viral diseases of birds than are such highly mobile creatures as mosquitoes and black flies. Mosquitoes and flies readily transmit disease-causing organisms from one bird to another because an individual mosquito or fly is likely to take blood meals from several different birds during its lifetime. By contrast, an individual mite—and even its descendants—will probably spend its entire life on the same bird. But many mites, especially internal parasites, are themselves the cause of diseases. Scabies of humans and other mammals, also known as the seven-year itch, is caused by nearly microscopic mites that burrow in the skin. Disease-causing mites of birds burrow in the skin or in the fatty layer just below the skin, and others cause serious symptoms or

even death by infesting the nasal cavities, the airways, the lungs, the air sacs, or even the digestive system or the body cavity in which the internal organs lie.

Ticks, really just big, blood-sucking mites, are divided into two groups: hard-bodied ticks, over 400 species with the upper side of the body more or less covered by a rigid shield; and soft-bodied ticks, less than 100 species that all lack a shield but have tough, leathery bodies. Soft ticks spend their lives in their host's nest, burrow, or other dwelling place, frequently visit the host at night to suck its blood, and mate and lay their eggs in or near the host's dwelling. Hard ticks are not associated with an animal or its dwelling except when they board a host to take a blood meal. Most of them require three different hosts to complete a life cycle that may last for more than a year. Tiny six-legged larvae hatch from eggs that were laid on the soil. They then climb up onto a blade of grass or some other low plant, board a passing bird or small mammal, drop to the ground after engorging, and then molt to the eight-legged nymphal stage. Nymphs board a second host, drop to the ground after engorging, and molt to the adult stage. Adults engorge on a third and last host, often a large animal, and lay their eggs on the soil after mating.

Many ticks attack mammals, some attack birds, and a few attack reptiles. Some species associate with either birds or mammals. The fowl tick *(Argus persicus)* is a soft tick that attacks only birds, usually fowl, both domestic and wild. (As is so often the case with all organisms, much more is known about this economically important species than about "unimportant" species that attack wild birds.) Fowl ticks hide in cracks and crevices during the day, but both immature forms and adults come out at night to suck blood from the chickens or other fowl with which they associate. Larvae hatch from eggs laid in the hiding places of the adult tick and remain firmly attached to a chicken or some other host as they feed for about five days. Then they crawl away from the host to a hiding place where they molt to the eight-legged nymphal stage. Unlike hard ticks, which have only one nymphal stage, fowl ticks and other soft ticks may have several, sometimes as many as four in the fowl tick. The nymphs feed on blood after each molt, including the final molt to the adult stage.

Another soft tick, a species of *Ornithodrus,* associates with both ground squirrels and burrowing owls. William Jellison found as many as 491 of these ticks per nesting burrow of these owls in the state of Washington. According to Jellison, and more recently Hal Harrison, burrowing owls often line their burrows and their nests with horse or cow manure. Jellison found many ticks in the dry manure. He showed that these ticks had fed on the owls rather than on nearby ground squirrels by crushing several of them and examining the blood in their guts under a microscope. The red blood cells had nuclei, proving that the blood that the ticks ingested did not come from the ground squirrel, a mammal. The red blood cells of mammals have no nuclei, while those of other vertebrates, including birds, do have nuclei.

The rabbit tick *Haemaphysalis leporispalustris,* a hard-bodied tick, occasionally takes blood from mice or birds such as quail and meadowlarks. In Minnesota, and surely elsewhere in the North, snowshoe hares are its most important hosts, but ruffed grouse are a close second. R. G. Green and his coworkers found as many as 12,000 of these ticks on a ruffed grouse and 16,000 on a snowshoe hare. When the population of hares is at low ebb, these ticks are largely dependent on the grouse. A closely related tick is rarely found on mammals but regularly attacks sage grouse and other chickenlike birds.

There are other hard ticks that associate with birds. A cosmopolitan species of *Ixodes,* known as the guillemot tick, has been found associated with guillemots, fulmars, gannets, puffins, and other sea birds that nest on sea islands or on narrow cliff ledges hundreds of feet above the sea. A related species parasitizes only cormorants.

Various species of flies are prominent among the insects that parasitize birds. Among them are louse flies, midges, black flies, nest flies, and mosquitoes. Miriam Rothschild and Theresa Clay put it nicely:

> If we could talk to birds as we talk to each other we would probably find that flies loom very large in their lives and provide one of the major topics of conversation. By day they form a favourite article of diet for many birds, but during the night the tables are turned with a vengeance. Incidentally it is an act of great cruelty to leave a canary uncovered in a cage after dark, for it is then assailed by all the female house-gnats [mos-

quitoes], which during the day, sit about silently on the walls and ceiling of the room.

Contrary to what many people think, not all mosquitoes drink blood. Males never do; they sip only exposed plant fluids, especially nectar from flowers. Even the females of a few species never take blood meals. Those of the genus *Toxorhynchites* drink only nectar and other exposed plant fluids. Adults of both sexes of the Asian and African genus *Harpagomyia* have the most bizarre feeding habits of any mosquito. They eat only the liquid stomach contents of certain ants, waylaying them on a branch of a tree and inducing them to regurgitate just as they would to feed a fellow ant. But most female mosquitoes are blood-thirsty creatures that obtain the protein they require to develop their eggs by taking blood from an animal. But like the males, they meet their requirement for energy by drinking the sugar-rich nectar from plants. The source of the blood meal is usually a warm-blooded vertebrate, a bird or a mammal. But sometimes a cold-blooded vertebrate such as a snake, turtle, lizard, frog, or toad may be the victim, and on rare occasions even an insect such as a cicada or a caterpillar.

Adult mosquitoes are choosy about the animals they take blood from, but none of them are as choosy as the fussier species of biting lice or mites. Mosquitoes, unlike some biting lice, will generally not starve to death in the absence of their preferred host. Some *Anopheles* mosquitoes that transmit human malaria prefer to feed on cattle but will feed on humans when cattle are not present. Similarly, some females of the genus *Culex* prefer the blood of birds but will sometimes take blood meals from humans or other mammals. Other mosquitoes prefer warm-blooded to cold-blooded animals or vice versa. For example, species of *Uranotaenia* prefer cold-blooded animals, often taking blood meals from snakes, frogs, or toads, but they will sometimes also feed from birds.

Mosquito bites cause people discomfort that usually involves redness and swelling at the site of the bite and an itching that invites vigorous scratching. These are largely reactions to the saliva that the mosquito injects when it pierces the skin to find blood. Birds bitten by mosquitoes probably suffer similar distress and scratch mosquito bites with their beaks or their feet. Mosquitoes can bite naked nestlings anywhere on their bodies, but usually

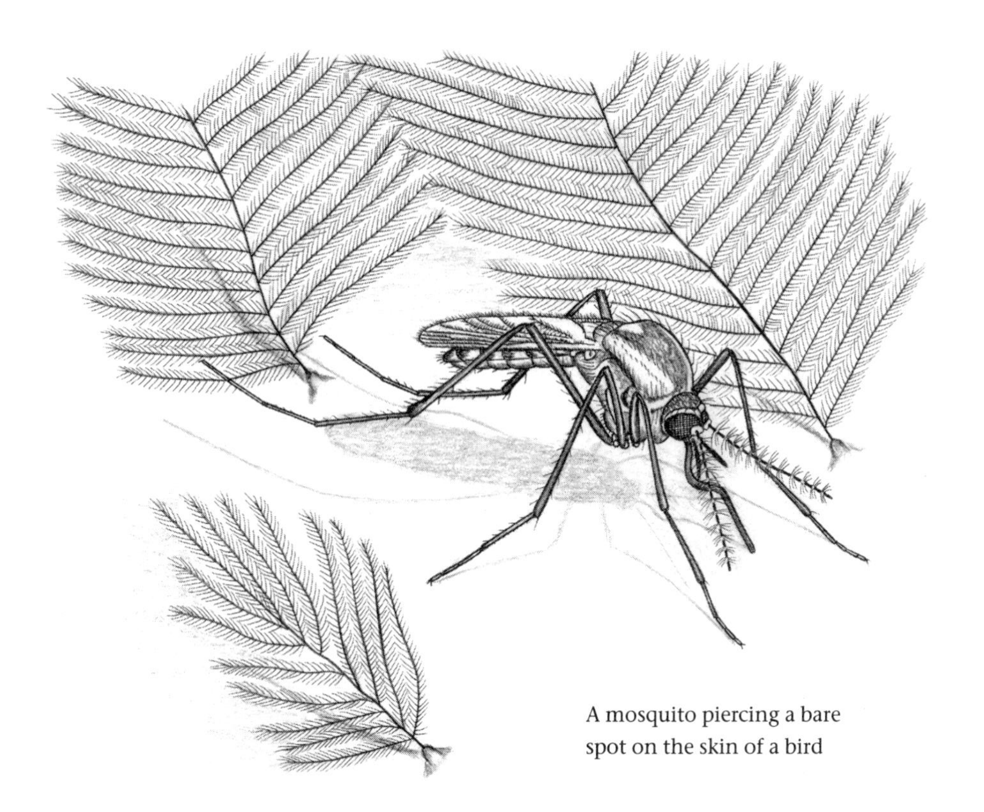

A mosquito piercing a bare spot on the skin of a bird

bite fully feathered birds on areas of bare skin such as combs, wattles, and the unfeathered areas of skin around the eyes.

But the bite of a mosquito can lead to far more serious consequences than relatively minor discomfort. It can also transmit disease-causing organisms. There immediately come to mind such mosquito-borne diseases of humans as yellow fever, caused by a virus; filariasis (its best-known symptom is elephantiasis), caused by a tiny nematode worm; and the malarias, caused by protozoa (single-celled creatures related to the amoebas) of the genus *Plasmodium.*

Birds also suffer from mosquito-borne diseases. They are susceptible to several forms of malaria that are caused by protozoa that are related to those that cause human malaria. Furthermore, they suffer from many non-malarial mosquito-borne diseases. For example, not only do birds harbor the mosquito-borne virus that causes eastern equine encephalitis in horses

and humans, but they themselves suffer severe symptoms and frequently die from the effects of this virus.

In the parlance of epidemiologists and medical entomologists, a zoonosis is a disease of both humans and other animals; arbovirus is shorthand for arthropod-borne virus; the reservoir of a disease of humans consists of the nonhuman animals in which the causative agent occurs; and a vector is the agent that transmits the disease-causing organism from animal to animal. Hence the disease called eastern equine encephalitis is a *zoonosis* caused by an *arbovirus* that is harbored in a *reservoir* that includes many different kinds of birds, horses, and perhaps other mammals or even reptiles, and that is transmitted by the many species of mosquitoes that are its *vectors,* especially various species of the genera *Culiseta* and *Culex.*

When eastern equine encephalitis is not in its epidemic (outbreak) phase, it is, of course, in its endemic (restricted) phase, confined to its "home base" in swampy areas. As C. D. Morris says, when the disease is in its endemic phase, mosquitoes circulate the virus among the birds of the swampy home base habitat: waders such as egrets, herons, and ibises; various shorebirds; owls; and many passerine species, including common grackles and red-winged blackbirds. From time to time, the virus moves out to the greater environment and into other species of birds, carried by mosquitoes that range beyond the swampy habitat. Eventually it may reach areas where humans or horses are present and be transmitted to the birds there, notably house sparrows and rock doves; such domestic species as chickens, ducks, and turkeys; and ring-necked pheasants and chukars that are being raised for release. In 1984, several captive whooping cranes in New Jersey died from eastern equine encephalitis.

Mosquitoes that take blood meals from both birds and mammals may transmit the virus from birds to horses or humans. Horses are frequently infected and often die from the disease. Fortunately, humans are seldom infected. There has been an average of fewer than 15 recorded cases per year since the eastern equine encephalitis virus was first isolated from humans in 1938. There are two forms of the disease in humans, one in which the brain and other parts of the central nervous system are infected and another in which the central nervous system is not involved. The latter form of the disease is generally not fatal and recovery is complete. But if the brain is infected, the mortality rate is high, about 70 percent, especially in children and the elderly, and there may be permanent brain damage in those who manage to survive.

Wild birds have been known to die from eastern equine encephalitis, but, as is to be expected, very little is known about the mortality rate in wild birds. We do know, however, that mortality in captive birds can be very high. For example, G. C. Parikh and two of his colleagues reported a 1967 outbreak of this disease on a South Dakota pheasant farm. The number of pheasants exposed to the virus, all from 16 to 18 weeks old, was 10,862. The great majority of these birds died, 89.8 percent, or 9,754. This may, of course, be an atypical mortality rate, because the pheasants were closely crowded together in their runs. Wild birds that are not so closely packed may have lower mortality rates.

Just as humans suffer from their own forms of malaria, caused by four different species of *Plasmodium,* birds are afflicted by malarias that are unique to them. Bird malarias, caused by various kinds of *Plasmodium,* are known to occur in many species—even penguins from South Africa and New Zealand—and may ultimately be found to occur in virtually all birds except for those that live in areas where there are no mosquitoes, such as those species of penguins that spend their entire lives on the continent of Antarctica or in the nearby seas. In North America, malaria-causing plasmodia have been found in such diverse species as the Canada goose, Forster's tern, ruffed grouse, eastern screech owl, and many songbird species, such as tree swallows, gray catbirds, red-winged blackbirds, and orange-crowned warblers.

Malaria is among the major causes of the extinction of many of the native land birds of the Hawaiian Islands, birds that occurred only on these islands. Especially grievous is the extinction of all but 21 of the more than 50 species of honeycreepers *(Drepanididae)* that still survived when the first Europeans arrived. This is a family that evolved on Hawaii and occurs nowhere else in the world. Today 14 of the remaining species are endangered.

Captain James Cook's shipboard surgeon and naturalist, William Anderson, was seriously ill when the expedition discovered the Sandwich (Hawaiian) Islands in January of 1778 and died shortly thereafter. Even so, specimens of about 16 species of birds were collected. Anderson noted that among the articles that the natives brought to trade were many skins of a small red and black bird, the iiwi, that were tied together in bunches of 20 or more. The Hawaiian chiefs wore luxurious, colorful cloaks, each made of

the feathers of as many as 10,000 birds. Such cloaks can even now be seen at the Bishop Museum in Honolulu.

One of several methods used to trap these birds requires patience and steely nerves. The trapper lay face up on the ground, his body covered with bushes and flowers, and he lightly gripped the base of a flower between thumb and forefinger. When a bird inserted its long, curved bill deep into the flower, the bill was quickly grabbed and the bird was pulled down beneath the camouflaging bushes.

You might think that the native Hawaiians were responsible for these extinctions. As you will read later on, they were responsible for many extinctions that occurred before the arrival of Europeans. But early-nineteenth-century European explorers and naturalists reported that many native birds were present and often were plentiful, despite the depredations of the Hawaiians. But by the late nineteenth century the land birds of the Hawaiian Islands were declining, obviously largely because of the activities of European and American settlers. This was partly the result of habitat destruction, the clearing of the native forests from the lowlands and the lower slopes of the mountains. But malaria and another mosquito-borne disease known as birdpox were also taking their toll. In 1902, H. W. Henshaw reported that "dead birds are . . . found rather frequently in the woods on the island of Hawaii, especially the iiwi and the akahani [= apapane]." He continued:

> The author has lived in Hawaii only six years, but within this time large areas of forest, which are yet scarcely touched by the axe save on the edges and except for a few trails, have become almost solitude. One may spend hours in them and not hear the note of a single native bird. Yet a few years ago these same areas were abundantly supplied with native birds, and the notes of the ou, amakihi, iiwi, akakani, omao, elepaio and others might have been heard on all sides. The ohia blossoms as freely as it used to and secretes abundant nectar for the iiwi, akakani and amakihi. The ieie still fruits, and offers its crimson spike of seeds, as of old, to the ou. So far as human eye can see, their old home offers to the birds practically all that it used to, but the birds themselves are no longer there.

We now know that bird malaria was largely responsible for the disappearance of these native birds. Malaria-causing plasmodia existed on the Hawaiian Islands long before these islands were discovered by Captain Cook. They were carried there in the bodies of ducks and shorebirds, such as the Pacific golden plover, sanderling, and ruddy turnstone, that migrated from North America to spend the winter in Hawaii. Even so, the plasmodia never spread to the native, nonmigratory birds of Hawaii. They remained locked in the bodies of their migratory hosts because there were no blood-sucking mosquitoes to transmit them to other birds. Mosquitoes of any sort were unknown in the Hawaiian Islands until they were unintentionally introduced. F. J. Halford wrote:

> Dr. Judd was called upon to treat a hitherto unknown kind of itch, inflicted by a new kind of *nalo* (fly) described as "singing in the ear." The itch had first been reported early in 1827 by Hawaiians who lived near pools of standing water and along streams back of Lahaina [a stopping place for whalers on the island of Maui]. To the Reverend William Richards, their descriptions of the flies suggested a pestiferous insect, from which heretofore the Islands were fortunately free. Inspection confirmed his fears. The mosquito had arrived!
>
> Investigation back-tracked the trail to the previous year and the ship *Wellington,* whose watering party had drained dregs alive with wrigglers [mosquito larvae] into a pure stream, and thereby to blot one more blessing from the Hawaii that had been Eden. Apparently no attempt was made to isolate and destroy the hatchery, nor to prevent spread of the pest throughout the archipelago. The pioneer was *Culex quinquefasciatus,* the night mosquito.

With the introduction of this mosquito, the stage was set for the transmission of malaria to Hawaiian land birds from the migrants from North America. Richard E. Warner, of the University of California at Berkeley, did experiments that prove the susceptibility of Hawaiian birds to this disease. He experimented with Laysan finches (actually drepanidids, not true finches) from the remote and isolated Hawaiian island of Laysan. This bird occurs only on Laysan and another remote island, and it had not been exposed to malaria, because there are no mosquitoes on these islands. Warner captured a number of Laysan finches and transported them to Kauai in cages wrapped with several layers of cheesecloth, thus protecting them from contact with mosquitoes or other insects. They were held in a mos-

quito-proof room for a month, during which time they remained healthy except for one bird that died. Then they were divided into two lots of 13 birds each. One group was exposed to mosquitoes by placing it in a coarsely screened cage in a shaded place out of doors; the other group remained in the mosquito-proof room. By the end of the sixteenth night of exposure, the birds in the outdoor cage had all died. Before they died, they showed signs of debility, and their blood contained massive infections of *Plasmodium.* All of the birds in the mosquito-proof room survived and showed neither symptoms of disease nor plasmodia in the blood.

Warner's experiment leaves little doubt that mosquitoes transmitted malaria to the birds kept in outdoor cages, and that the Laysan finch is particularly susceptible to this disease. Probably all of the native Hawaiian birds are especially susceptible to malaria. Since they evolved in the absence of mosquitoes, they were never exposed to this disease and did not develop resistance to it.

In an article in the September 1995 issue of *National Geographic* Elizabeth Royte chronicled the current plight of the native Hawaiian flora and fauna. She relates that birds literally fell from the trees during recent outbreaks of avian malaria, and explains how feral pigs exacerbate the malaria problem in the few remaining forests of native vegetation. At least 100,000 wild pigs roam the islands, descendants of the large swine introduced by Europeans and the small domestic pigs brought by the Polynesians, who first settled these islands about 1,400 years ago. The pigs knock over giant tree ferns and hollow them out to eat the starchy core. The pools of rainwater that gather in these hollow trunks serve as important breeding sites for the larvae of the mosquitoes that transmit bird malaria.

✱ Among the other flies that are blood-feeding parasites of birds are the black flies. You will hear more about them later when I tell you about bugs that eat people. Although these vicious blood suckers often inflict their painful bites on mammals, they also attack birds, and in so doing may transmit disease-causing organisms. One North American species transmits a malaria-like protozoan of ducks, and another, known as the turkey gnat, transmits a similar disease-causing parasite of turkeys.

Certain kinds of tiny midges (family Ceratopogonidae) also attack birds.

*Above:* 1. A blissful fledgling screech owl about to gulp down a cecropia moth

*Below:* 2. During the day, nocturnal tiger moths show off their colors to warn birds of their toxicity

*Above:* 3. A solicitous male cardinal feeding his mate a caterpillar

*Below:* 4. Soldier beetles mating

*Right:* 5. A male mountain bluebird pausing before bringing a cricket to its nestlings

*Below left:* 6. A long-horned beetle that feeds on milkweeds is warningly colored

*Below right:* 7. A monarch butterfly stopping to take a drink on its way south

8. A female American kestrel delivering an insect to an almost fully grown nestling

*Above:* 9. A goliath bird-eating spider of South America devouring a nestling bird

*Right:* 10. This harmless robber fly scares off birds by mimicking a stinging bumble bee

11. An amorous pair of six-spotted tiger bettles

*Above:* 12. A giant swallowtail caterpillar that looks like a bird dropping prepares to molt to the pupal stage

*Right:* 13. Can you find the camouflaged inchworm moth in this photograph?

Hint: it's just to the left of center

14. This planthopper passes itself off as a leaf

*Above:* 15. Mr. Personality, a spicebush swallowtail caterpillar whose thorax looks like a big head

*Below:* 16. This dragonfly, a skimmer, will dart out to snatch insects from the air

*Above:* 17. A male firefly flashing to a potential mate as he flies in the dark

*Center:* 18. This harmless syrphid fly mimics a stinging yellowjacket wasp

*Below:* 19. This wheel bug, a fierce predator of insects, will bite if handled

*Above:* 20. An eastern bluebird taking an exuberant bath that will discourage some of its parasites

*Below:* 21. Bobwhite quail taking dust baths that will help rid them of lice

*Above:* 22. The underside of a well-camouflaged gray comma butterfly

*Below:* 23. A regal fritillary sipping nectar from butterfly-weed, an orange milkweed

*Above:* 24. This female black swallowtail was once a parsleyworm caterpillar

*Center:* 25. This pipevine swallowtail caterpillar's colors warn of its noxiousness to birds

*Below:* 26. Still soft and fragile, an adult cicada emerges from its nymphal skin

*Left:* 27. This hickory horned devil will become a regal moth

*Right:* 28. This cactus wren, a caterpillar in its beak, is approaching its nest in a cholla

*Above:* 29. This zebra swallowtail's caterpillar offspring will eat only pawpaw leaves

*Center:* 30. A female tiger swallowtail of the non-mimetic form

*Below:* 31. This io moth's large eyespots, suddenly revealed, are probably enough to startle any bird

*Above:* 32. This pair of cecropia moths will stay coupled for about 15 hours

*Center:* 33. A male promethea moth perched on the cocoon from which it emerged

*Below:* 34. A Bell's vireo feeding a caterpillar to a nestling

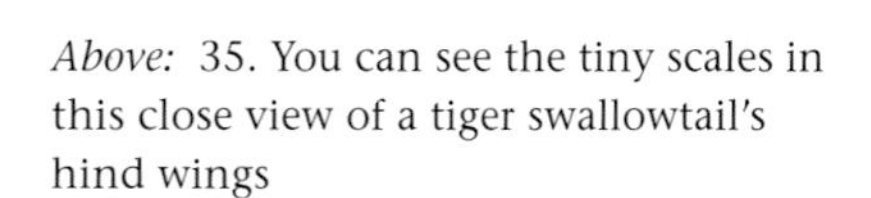

*Above:* 35. You can see the tiny scales in this close view of a tiger swallowtail's hind wings

*Center:* 36. This gray hairstreak has a false head complete with antennae on its hind wings

*Below:* 37. The business end of a Chinese mantis ready to strike at its insect prey

38. A chipping sparrow feeding a mantis to a nestling

Most of the flies of this family suck the juices of plants or capture small insects and suck their bodies dry. Others, parasites rather than predators, cling tenaciously to the wing of an insect, often a large moth, a dragonfly, or a green lacewing, as they suck blood from one of the veins. A few species suck blood from birds, humans, and other vertebrates. The bites of the species that attack humans, commonly known as biting midges or no-see-ums, are very painful and irritating. Indeed, as Miriam Rothschild and Theresa Clay noted, it has been jokingly conjectured that the presence of an exceedingly pestiferous species in Scotland—in conjunction with the wearing of the kilt—gave rise to the Highland fling. At least one of these midges gets its blood second-hand. It lands on the body of a mosquito that has just engorged itself with blood—perhaps that of a bird—and pierces the mosquito's abdomen to steal a meal.

Only a few kinds of biting midges are known to attack birds, but there may be several more species that do so, since the feeding behavior of many of these insects is still not known. In North America, investigators have found large numbers of one species engorged with blood in the nests of magpies and crows. Some species are known to transmit malaria-like diseases of birds such as ducks and spruce grouse.

✱ Shiny blue flies of the genus *Protocalliphora,* often called nest flies, drink nectar from flowers and look like other blue-bottle flies, but, unlike the other members of this family, when they are in the maggot stage they are blood-thirsty parasites of nestling birds. Worldwide, there are several genera of these flies. But a dozen or more species of *Protocalliphora* are the only North American representatives and the only North American flies of any sort that suck blood as larvae.

Adult females, apparently attracted by the odor of nestling birds, lay a small number of eggs—often only 10 to 20—in nests occupied by partly grown nestlings. They do not lay in empty nests or those that contain only eggs. The sluglike maggots hatch within a day and immediately burrow into the nesting material. Much like leeches, hungry maggots crawl onto the nestlings, usually at night, and attach themselves by means of a sucker at the front end of the body as they drink blood for about an hour before leaving the bird to burrow back into the nest. The maggot, after taking

several blood meals over a period of about 12 days, completes its growth and transforms to the pupa in the nesting material. The adult flies emerge 2 or 3 weeks later.

It seems that all, or almost all, North American birds that build nests are attacked by one or another species of nest fly. These flies do not seem to be host specific, although a species may have habitat preferences—perhaps for cavity nests rather than cup nests or vice versa. As Catherine A. Rogers and two of her colleagues reported in 1991, a nest fly that infests tree swallow nests in Ontario has also been found associated with several other cavity-nesting birds, including other swallows, screech owls, downy woodpeckers, great crested flycatchers, house wrens, European starlings, and eastern bluebirds. Nest flies, which also occur in Eurasia, have been found in the nests of many British birds, including nightingales, redstarts, skylarks, meadow pipits, and various tits, wagtails, crows, swallows, and martins.

Infestations of nest flies are not uncommon. The majority of tree swallow nests seem to be infested: 82 percent of those examined in one study and 72 percent in another study. According to Clifford Gold and Donald Dahlsten, of 127 nests of the mountain chickadee examined in Modoc County, California, almost 93 percent contained anywhere from 1 to over 100 nest fly maggots. Similar results were obtained when they examined nests of the chestnut-backed chickadee in El Dorado County, California.

Individual nests are often infested by 10 or 20 maggots, but the number may be much higher, as many as 373 in one nest of a black-billed magpie in Europe. In Ontario, 167 were found in a tree swallow nest that held only three nestlings. In California, a chestnut-backed chickadee nest contained 273 maggots, and almost 13 percent of the mountain and chestnut-backed chickadee nests examined contained 100 or more maggots.

There has been controversy about the effect of these maggots on nestlings. Some early researchers reported that they may severely weaken nestlings and often cause their death, but recent reports based on systematically collected data from large numbers of nests indicate that the great majority of nestlings survive to leave the nest. In a study of 335 tree swallow nests, Catherine Rogers and her colleagues found that even "heavily parasitized nestlings do not necessarily suffer unusually low growth rates or high mortality." Nevertheless, the loss of blood by a nestling may be very great. Perhaps the nestling is able to make up for the loss, particularly since it is gradual—occurring over a period of about two weeks. But even if a nestling is able to leave the nest, it may be so weakened that it cannot survive the post-fledging period. However, data on the survival of fledged birds from infested nests are not available and will be difficult to obtain.

The louse flies (Hippoboscidae) constitute a small but geographically widespread family of blood feeders, most of which attack birds but some of which attack mammals. These flattened, leathery, sparsely hairy creatures—often considered to be disgusting in appearance even by entomologists—vary widely in the degree of their commitment to the animals that they parasitize. Some species are fully winged throughout their adult lives and fly from one host to another. Others are winged but shed their wings once they have located and boarded the host on which they will spend the rest of their lives. A few species that lack wings or have only vestigial wings also spend their entire adult lives on one host individual.

Louse flies do not have a free-living larval stage. Instead, the female retains the maggot, or larva, within her body, not giving birth to it until it is ready to metamorphose to the pupal stage. This process is strikingly like mammalian reproduction. The oviduct of the female is enlarged to form a "uterus" that holds the developing larva, and the larva is nourished by a gland that secretes milk into the uterus. This insect milk is similar in appearance and chemical composition to the milk of humans and other mammals.

The newly born maggot prepares for pupation in the unique manner of all of the "higher" flies—those other than the mosquitoes, crane flies, midges, and gnats. Just before or just after it is born, the larva assumes an oval shape and darkens and hardens its soft white skin. Only then does the pupa separate itself from its larval skin, but it does not shed it. It remains within it, protected by the newly hardened larval skin as the pupa of a moth is protected by its silken cocoon. After the pupa matures and the adult fly molts its pupal skin, the fly emerges from the puparium, as the hardened larval skin is known, by bursting its way out with an inflatable bladder that protrudes through an opening at the front of the head. After emergence, the bladder is retracted and the opening grows shut.

The formation of the puparium and the retention of the larva in the

uterus have made it possible for some mammal-infesting louse flies, notably the sheep ked, to spend their entire lives on one host animal. The wingless adult, sometimes inappropriately referred to as the sheep tick, crawls about like an oversized louse in the fleece, moving to another sheep only if the latter comes into close contact with the ked's original host. When a female ked gives birth to her offspring, a full-grown maggot, she glues it to the fleece, where it will immediately form a puparium and remain firmly fixed until it metamorphoses to the adult stage and emerges to take up its life as a permanent, blood-feeding ectoparasite of the sheep.

Among the louse flies that infest birds, the disposition of the puparium differs with the nesting habits of the host. Many of the louse flies that parasitize birds that reuse the same nest year after year—such as swifts and bank swallows (called sand martins in Europe)—place their puparia in the host's nest. Here the pupal louse fly spends the winter sheltered by the nest and the puparium. In the spring the adult fly emerges from the puparium and makes its way to one of the birds that returns to reuse the nest. Many louse flies that infest birds that reuse nests have very small or otherwise aberrant wings that are useless for flying. They have no need to fly in search of a host; under most circumstances, a host will come back to them. But the louse flies that infest birds that build a new nest every year have wings because they must fly in order to locate a new host. These parasites deposit their puparia on the ground, where they are hidden by fallen leaves and other organic debris during the winter.

During the summers of 1957 to 1960, Gordon F. Bennett studied three species of louse flies that occur in Algonquin Provincial Park in Ontario. Of the 6,448 birds of 84 species that he captured and examined, about 16 percent were infested with louse flies, most of them by *Ornithomyia fringillina,* a winged louse fly that attacks not only fringillids such as sparrows and evening grosbeaks, but also warblers, thrushes, and a variety of other songbirds, and occasionally even saw-whet owls and woodpeckers.

Although their host range is wide, flies of this species did show certain preferences. They were far more likely to be found on immature birds than on adults, and on thrushes, warblers, and sparrows than on other birds. They would, however, rather starve to death than drink the blood of purple finches. All 41 of them that had been artificially transferred to purple finches died within four days. But flies transferred to white-throated spar-

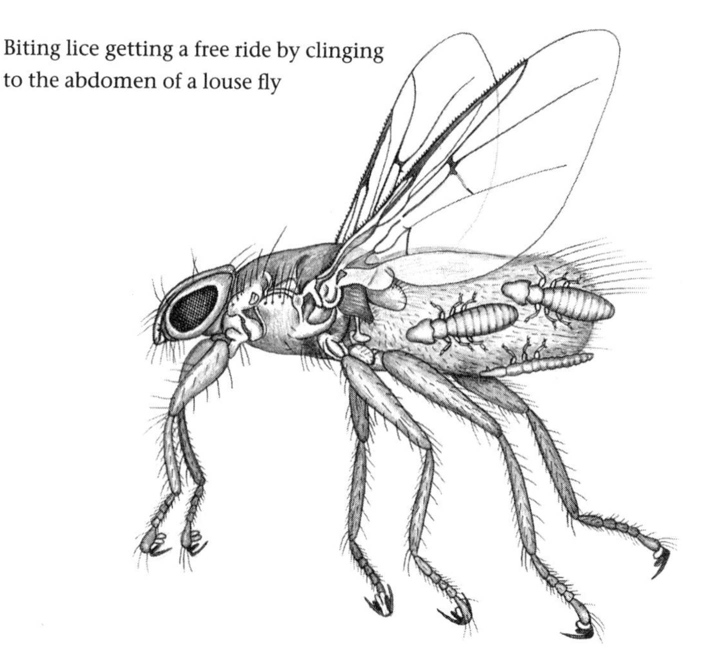
Biting lice getting a free ride by clinging to the abdomen of a louse fly

rows survived. *Ornithomyia fringillina* was more likely to be found on birds that nest on or near the ground than on those that nest in trees at medium to high levels. Among the warblers, for example, 16 percent of the low-nesting birds were infested, as compared to only 5 percent of those that nest at medium to high levels. As Bennett pointed out, louse flies, such as *Ornithomyia fringillina,* that emerge from puparia on the ground are likely to first encounter birds that frequently are near the ground.

Some louse flies unwittingly provide transportation for lice seeking a new host. In 1975, in the precise but dry language of the biologist and without any hint of anthropomorphism, James Keirans offered the following definition of this association between fly and louse:

> Phoresy is a transitory association between 2 species in which one animal attaches to another and is carried away from a potentially suboptimal environment. In the case of Mallophaga [the biting lice], a smaller biting louse attaches by its mandibles to a larger hippoboscid, or louse fly, and is transported away from a presumed less favorable environment, usually a dead bird. Shelter is afforded the mallophagan, but it does not feed on the

> hippoboscid nor, if an immature stage, exhibit ontogeny [development] while attached.

Phoresy has been called biological hitchhiking, but this term is not quite appropriate. Hopping a freight is more like it. A phoretic passenger is more like a hobo stealing a ride on a freight train than a hitchhiker who stands by the side of the road asking for a lift. As far as we know, the louse does not ask permission; it just climbs onto the louse fly and holds on. If the louse fly lands on a living bird, the louse will climb off the fly and make itself at home if the bird is a suitable host.

Phoresy is not an uncommon phenomenon. Considering only biting lice and louse flies, at least 405 instances of phoresy have been recorded in the scientific literature. These passenger-carrying flies had been collected from birds of 89 different genera from 33 different families, ranging from bitterns and egrets to sparrows and orioles. A significant percentage of louse flies carry lice as passengers. Figures cited in the literature include 22 percent of a sample of 180 louse flies, 43.5 percent of another sample of 156, and 25.5 percent of yet another sample of 55. Most louse flies carry only one or two lice, but in one instance a single louse fly had 31 lice on board.

As Miriam Rothschild and Theresa Clay wrote:

> The simplest way to collect bird fleas is to take a nest from which the fledgelings have recently flown and to keep it in a cardboard box or linen bag. Providing the nest is damped periodically, the larval or pupal fleas continue to develop in the debris or rubbish in the bottom, and in due course hatch out. It is a more lengthy process to collect them off the bodies of their host. Less than one bird in ten harbours fleas, and then generally only one or two specimens at a time. Moreover the host has to be enclosed in a receptacle immediately after being shot or captured, otherwise the fleas hop off and escape. The maximum number recovered from a bird is 25 specimens from a house-martin, a species of swallow that does not occur in the Western Hemisphere. On the other hand, no less than 4,000 have been bred out of a single martin's nest.

This is good advice for those of you who might want to capture a few fleas, and it is also a revealing observation on the life history and habits of these wingless, blood-sucking parasites of mammals and birds, which live in the nests of birds as larvae and as adults visit the resident birds to feed.

Fleas are among the few parasitic insects with complete metamorphosis that are in one way or another associated with the host throughout both their adult and larval lives. As adults, they use their piercing and sucking mouthparts to take blood meals from the host's body; some species make occasional brief visits to the host, others may spend long periods of time on the host, and a few live more or less permanently on the body of the host. Among the last group are the adult female of the chigoe flea (not to be confused with the chigger), which burrows under the skin of a human or other mammal, and the adult female sticktight flea, a parasite of birds that buries its head in the skin of the host's head and can be found in the southern United States on wild birds or domestic poultry, especially on the combs and wattles of chickens.

Many fleas lay their eggs in the nest, den, or other sleeping place of the host. Some lay them on the body of the host, but these eggs eventually fall to the floor of the nest or den. The female sticktight flea remains attached to the host, but expels her eggs forcibly from her insectan analogue of a vagina, propelling them for some distance and thus assuring that they will fall from the body of the bird and, with luck, land in its nest or roosting place. The larvae that hatch from flea eggs resemble tiny caterpillars but are eyeless and legless. Each larva uses its chewing mouthparts to eat organic debris in the nest, including dried droplets of blood that have passed virtually undigested through the intestines of an adult flea—perhaps the larva's own mother or father—and fallen from its body.

Full-grown larvae pupate in tiny silken cocoons that are usually festooned with bits of debris. In many species, including fleas of birds, the adult delays its emergence from the cocoon until some cue informs it that a host is actually present in the nest or den. Although the mature adult sheds its pupal skin early on, it does not emerge from the cocoon until vibrations or perhaps an increase in temperature or an elevated carbon dioxide level signals the presence of a potential host. This behavior is adaptive, increasing the likelihood that the flea will survive. If the host is absent—often for months at a time—there will be nothing for the newly emerged flea to eat. It will be better off to remain quiescent in the cocoon and thus conserve its energy until a host appears. The late Gottfried Fraenkel, director of my Ph.D. research and one of the world's great insect physiologists, demon-

strated the fleas' response to outside cues to his classes by banging on a table on which he had placed a covered dish containing cocoons of the oriental rat flea. The hungry fleas, previously undisturbed for weeks, responded almost immediately to the vibrations, which they took as indicating the arrival of a host, by emerging from their cocoons and jumping about frantically in search of a host that was not there.

Almost all fleas associate with hosts that have a permanent home where the larvae can live and feed—such as the nest of a bird or the den of a wolf or some other mammal. The continued association of the host with its home ensures that the fleas will have a source of blood meals after they metamorphose from the larval to the adult stage. (Note that humans, the only primates that keep a permanent home, are the only primates that are infested by fleas.) But, as is almost always the case in matters biological, there are exceptions to this generalization. As Adrian Marshall points out, the larvae of certain fleas that parasitize hoofed mammals live on the ground where their hosts pasture. Fleas of the species *Uropsylla tasmanica* are at the other end of the spectrum. They are probably the only fleas that are parasites of the host during the larval stage. In Tasmania and Australia these anatomically specialized larvae burrow into and feed on the skin of the marsupials that are their hosts.

The 1991 edition of the *Insects of Australia*, compiled by the Division of Entomology, Commonwealth Scientific and Industrial Research Organization, estimates that there are about 2,380 known species of fleas in the world. The great majority of them are parasites of mammals, most of them associated with rodents (about 74 percent). Only 132 species are associated with birds. The number of bird species attacked is far greater than the number of bird fleas. Although some bird fleas are more or less specific in their choice of hosts, others are catholic in their tastes. In Britain, for example, the hen flea attacks at least 75 different species of birds.

The life cycles of the fleas of birds are closely linked with the nesting activities of their host. The closeness of this link and the nature of the behavioral mechanisms that mediate it are exemplified by the relationship between the bank swallow and the flea that parasitizes it. In Europe, where you will remember it is called the sand martin, it is parasitized by a flea called *Ceratophyllus styx,* and in North America it is parasitized by *Ceratophyllus riparia*. Robert Lewis, one of the foremost experts on fleas, recently told me that these two fleas really belong to the same species.

Bank swallows, which are highly colonial, raise their offspring in a nest of grass and feathers placed at the end of a long burrow that they excavate in a steep-sided, sandy bank, often in a streamside bluff or the clifflike face of a sand pit. The North American population overwinters in South America, and the Eurasian population in either Africa or southern Asia. In spring these swallows return to the same nesting site they occupied the previous year, sometimes digging new burrows but more often reusing old ones.

The fleas of bank swallows are highly host specific and almost never occur on other birds. They are not present on the birds in winter, but remain behind to overwinter in large numbers in the temporarily unoccupied nests, almost all of them as mature adults poised to emerge from their cocoons at a moment's notice. The next spring, after being stimulated to emerge by some disturbance of the nest, normally caused by a returning swallow, they either board the returning bird or move to the burrow's entrance, often forming large aggregations there, and attempt to board any swallow that approaches the nest.

In a series of simple but beautifully designed experiments, David A. Humphries of the University of Aston in Birmingham, England, elucidated the behaviors, and their underlying mechanisms, that enable bank swallow fleas to exploit their hosts. They must become active when their hosts return in the spring; they must find mates; and they must disperse to other burrows if their burrow is not reoccupied. Humphries's results show that the behavior of these fleas can be explained in terms of simple, instinctive mechanisms. Thus it is neither necessary nor realistic to assume that the flea, with its minimal brain, has a goal in mind or that it plots to achieve that goal.

The first task that the fleas must accomplish in spring is to coordinate their emergence from the safety of the cocoon with the arrival of the returning swallows. Humphries reasoned that the fleas must become active either in response to the presence of a swallow that disturbs the nest or to a change in some climatic factor, possibly the seasonal increase in ambient temperatures, that happens to coincide with the arrival of these birds. He then did experiments to determine which of these factors is actually responsible for activating the fleas. One experiment, done in late autumn after the swallows had left, showed that a mechanical stimulus is sufficient to activate bank swallow fleas. He disturbed the nests in several burrows that had been occupied the previous summer "with a long flexible cane

bearing on its tip a stiff tassle of rope." Within two days fleas began to appear at the entrances to these burrows but, as he expected, none appeared at similar burrows that had not been disturbed. Another experiment, done in spring, eliminated climatic factors from consideration. Before the swallows arrived he used thin twigs to block the entrances to several burrows that had been occupied the previous spring. Birds could not enter but fleas could exit. After the swallows returned, many fleas emerged from open burrows that had been entered by swallows, but none appeared at blocked burrows. The obvious conclusion is that the fleas emerged from their cocoons in response to disturbances by swallows, and that the increasingly warm temperatures of spring, which certainly can be felt through a few thin twigs, had no discernible effect. Only a brief disturbance is necessary. A swallow that examines the nest and then leaves activates the fleas as effectively as a swallow that stays on to occupy the nest.

Humphries next sought an explanation for the movement of these fleas to the entrance of the burrow, which is usually from 1.5 to 3.0 feet away from the nest in which they spent the winter. He hypothesized that they were responding to the light at the opening of the burrow. Such responses are often phototactic, instinctive orientations to light. (Phototaxis may be positive, orientation *to* light, or negative, orientation *away* from light.) In laboratory experiments, he showed that fleas do indeed respond phototactically. Mature fleas, confined in a horizontal cardboard tube with a light at one end, moved toward the lighted end of the tube but reversed course whenever the light was shifted to the other end of the tube.

Other laboratory experiments showed that young adult bank swallow fleas are negatively phototactic for the first day after they emerge from the cocoon, but that they are positively phototactic thereafter. Why this shift in response to light? In nature, the brief initial period of negative phototaxis keeps the fleas within the nest long enough to mate with other newly emerged fleas or to make contact with a returning swallow that might have been temporarily absent. If the burrow remains unoccupied, the switch to positive phototaxis leads the flea to the entrance, where it has a good chance of boarding an exploring swallow.

Bank swallows frequently hover for as long as 30 seconds within 3 or 4 inches of a burrow entrance. They may then just fly away, or they may enter the burrow to explore. Even a swallow that only hovers in front of the burrow and then leaves can serve to disperse the fleas to other, perhaps occupied, burrows. Fleas that aggregate at the entrance to the burrow use their powerful legs to leap onto any bird that hovers nearby. As shown by Humphries's experiments, the instinctive responses of the fleas, honed to a fine edge by natural selection, are exquisitely suited to their dispersal via swallows that only hover in front of the burrow.

Humphries was first alerted to the dispersal behavior of bank swallow fleas when he saw many of them that were sitting at burrow entrances leap out into space when he cast a shadow upon them. He then did an experiment in which he used a small, square, black card to mimic the hovering of a bird in front of a burrow. When the card, mounted on a stiff wire, was held in front of several burrows, 85 of 103 fleas leaped to intercept it. In control experiments using the wire but no card, not one flea jumped out of 7 aggregations of fleas totaling 88 individuals. Clearly, the fleas were stimulated to leap by the decrease in light intensity caused by the presence of a card, and they would presumably be similarly stimulated by a hovering bank swallow. Their behavior is even programmed to avoid a futile response to bank swallows or other birds whose shadows may only briefly fall on the fleas as the bird quickly flits past without pausing. The fleas delay their response. They do not jump until the visual stimulus has been present for from 1 to 6 seconds.

Humphries then did a similar experiment, this time using a more or less realistic model of a bank swallow, to explore the effect of the distance of the bird from the burrow entrance. When the model, a piece of hardboard cut to the size and shape of a swallow's silhouette, was held about 3 inches from some fleas, 54 of them, 72 percent of 75 individuals, leaped to intercept it, and 39 of them actually hit the model. When it was held about 6 inches from the fleas, only 60 percent of 90 individuals leaped, and only one of them managed to hit the model. Fleas that leap but miss a bird fall to the base of the sand cliff. But they are not doomed. They immediately crawl up the cliff and position themselves for another leap from the entrance to a burrow.

✱ In addition to blood suckers and other parasites, there are a few insects and spiders that do occasionally eat birds in the manner of a carnivore,

as jaguars or ladybird beetles kill their prey outright and then devour it. Among them are some ants, praying mantises, and certain large spiders.

A few species of ants are the only insects that I know of that regularly kill and devour birds or other vertebrates. This is, of course, a cooperative effort of many ants from the same colony; few ants are so venomous that the sting of only one individual is likely to kill an animal as large as a bird. All ants are social, and all live in colonies that include a fertile queen, who does all the egg laying, and a number of sterile females, often many thousands of them, that make up the worker caste. It is, of course, the workers that go forth to forage for prey that will be carried back to the colony.

Among these predatory ants are the fire ants of North and South America, several species of the genus *Solenopsis,* which derive their common name from their painful, venomous stings. Bernard V. Travis, of the no longer extant Bureau of Entomology and Plant Quarantine of the U.S. Department of Agriculture, first reported, in 1938, that a fire ant native to Florida is a regular predator of bobwhite chicks, although its diet consists principally of insects and other arthropods. These ants sometimes kill newly hatched chicks, but they more often enter an egg that has just been pipped and sting the chick to death before it can break its way out of the shell. The flesh of the unfortunate chick is then torn to pieces and carried back to the colony. Travis reported that of 2,456 bobwhite nests observed from 1924 to 1937, 151 were attacked by this ant, an average of 6.2 percent per year. In 1928, a high of over 12 percent of the nests observed had had some or all of the chicks destroyed by fire ants. Another species of fire ant is known to destroy young California quail just when the eggs are pipped, and unidentified ants, possibly fire ants, attack Gambel's and scaled quail in Arizona.

In addition to the native species, there are two other fire ants that were unintentionally introduced into the United States from South America—one around 1918 and another in the 1940s. These two imported species also attack birds and other small vertebrates. The first introduced species is still confined to parts of Alabama and Mississippi, but the second one has spread through most of the southern United States.

The nomadic army ants of South America and the driver ants of Africa are the most notorious of the carnivorous insects that are known to attack and kill birds. In horror stories, army ants have been portrayed as threatening entire human settlements as they advance on a front that is miles wide.

These are, of course, wild exaggerations, but even when seen in its true and more modest dimensions, a raid by army ants is very impressive. The ants, anywhere from 150,000 to 700,000 strong, advance on a front that is often 60 feet wide.

The army ants that I discussed in an earlier chapter do not live in permanent nests as do other ants. They instead form temporary bivouacs in which the queen and the larval brood are sheltered only by the living bodies of a large and thick mass of worker ants that may hang from the bole of a fallen tree, nestle between the buttresses of a standing tree, or lie in some other protected place. A raiding swarm moves out from the bivouac early in the morning. The late T. C. Schneirla of the American Museum of Natural History in New York wrote of a swarm of army ants:

> The huge sorties of *burchelli* [*Eciton burchelli,* a species of army ant] in particular bring disaster to practically all animal life that lies in their path and fails to escape. Their normal bag includes tarantulas, scorpions, beetles, roaches, grasshoppers, and the adults and broods of other ants and many forest insects; few evade the dragnet. I have seen snakes, lizards, and nestling birds killed on various occasions; undoubtedly a larger vertebrate which, because of injury or for some other reason, could not run off, would be killed by stinging or asphyxiation. But lacking a cutting or shearing edge on their mandibles, unlike their African relatives the "driver ants" these tropical American swarmers cannot tear down their occasional vertebrate victims.

The driver ants of Africa—so named because they drive all living creatures before them—are related to and similar in some aspects of their behavior to the army ants. The ants of both groups are predators that send out large columns of aggressive workers that beat the bushes to ferret out prey animals that they kill and carry back to the colony. But while army ants bivouac above ground, driver ants dig subterranean nests. Furthermore, colonies of driver ants are usually much larger than colonies of army ants, consisting of as many as 20 million individuals. Both army and driver ants are nomadic to varying degrees; they must necessarily move to new hunting grounds after they have stripped an area of prey. Although, as Schneirla reported, the mandibles of army ants lack a cutting edge and are thus not suitable for cutting up the flesh of large prey animals, the driver ants have

sharp, serrated mandibles, with which they cut their prey into small pieces, even birds as large as domestic fowl.

In an 1845 letter to J. O. Westwood, Esq., secretary of the Entomological Society of London, the Reverend Thomas S. Savage, M.D., presumably a medical missionary, described the activities of driver ants in West Africa. The letter, published in the 1847 volume of the *Transactions* of the Entomological Society, reported that the driver ants "fiercely attack any thing that comes in their way, 'conquer or die' is their motto . . . They are decidedly carnivorous in their propensities. Fresh meat of all kinds is their favorite food":

> They attack . . . lizards, snakes, etc. with complete success. We have lost several animals by them . . . monkeys, pigs, fowls, etc. The severity of their bite, increased to great intensity by vast numbers, it is impossible to conceive. We may easily believe that it would prove fatal to almost any animal in confinement. They have been known to destroy the *Python natalensis,* our largest serpent. When gorged with prey it lies powerless for days; then, monster as it is, it easily becomes their victim.

Savage goes on to describe how drivers "butchered" a fowl they had killed:

> The feathers were *pulled out,* sometimes one, two and three ants would be seen tugging most lustily at one, but I am inclined to think that the largest feathers were extracted by lacerating the flesh at their root, though I was not able to decide this point fully . . . The operation of *picking* [cutting up the prey] began at the beak, and was gradually extended backward. The neck being half stripped, they then began the work of laceration at the eyes and ears. It was some time before any visible impression was made, but at last . . . deep cavities appeared, and muscles, membranes and tendons were reduced and borne off to their habitation. The juices and a portion of the muscular fibre, I think, must be consumed on the spot at such times, though the largest portion is carried to the domicil.

Praying mantises are among the very few predaceous insects that are large enough to eat a bird, but they have seldom been seen to do so. However, the December 1994 issue of *Birding* contained two letters to the editor that reported just such occurrences. Ralph J. Fisher, Jr., of Silver City, New Mexico, presented a description and a photograph of a large tan-colored mantis eating what Fisher identified as a young rufous hummingbird that it had caught in its raptorial front legs as it waited in ambush on a hummingbird feeder. In the previous year he had seen a large green mantis sitting on a feeder strike at but miss a rufous hummingbird. He also mentioned two other documented instances of mantises catching hummingbirds at feeders. Yet another case was reported in the same issue of *Birding* by Connie Gottlund of Cedar Creek, Texas. She noted that a praying mantis was seen eating a male ruby-throated hummingbird that it had caught at a feeder.

Mantises may also catch an occasional hummingbird under natural conditions. They often lurk near flowers, apparently in the hope of ambushing a nectar-seeking insect for a meal. A hummingbird that visits a flower to drink nectar may thus be caught just as is a hummingbird that visits a feeder to drink a solution of sugar in water.

Many people have heard of bird-eating spiders, but most of them know little about these spiders except for their bird-eating reputation. The very thought of a spider large enough to capture and eat a bird titillates the mind, but the story becomes even more fascinating when the facts are known. These large spiders are mainly nocturnal, usually spending the daylight hours in some shaded and protected spot. At night they venture out to run over the ground and the trunks and branches of trees in search of their prey. Although their diet consists mainly of insects and other arthropods, they take advantage of opportunities to kill and eat small vertebrates, sometimes birds that they take unawares in the dark.

What is probably the earliest report of a spider eating a bird was published in 1705 by Maria Sibylla Merian, a resident of the Netherlands colony of Suriname (Dutch Guiana) in northeastern South America from 1699 to 1701. Her book, with 60 hand-colored plates and texts in both Latin and Dutch, was printed in an edition of less than 100 copies. It is, of course, difficult to obtain and very valuable. But in 1982 an edition with texts in Dutch and English was published in the Netherlands. On her plate 18 she shows a large spider dragging a hummingbird from a nest that holds four

eggs, an unusually large number for a hummingbird. She wrote that spiders of this species feed mainly on the large ants that are shown on this plate, but that "if ants as food are lacking, these giant and dangerous spiders are able to drag small birds from their nests, and to kill them by sucking their blood."

Most naturalists discredited Merian's report until bird eating by a spider was again reported, in 1863, by Henry W. Bates in his book *Naturalist on the River Amazons*. Bates wrote:

> The [spider] was nearly two inches in length of body, but the legs expanded seven inches, and the entire body and legs were covered with coarse grey and reddish hairs. I was attracted by a movement of the monster on a tree-trunk; it was close beneath a deep crevice in the tree, across which was stretched a dense white web. The lower part of the web was broken, and two small birds, finches, were entangled in the pieces; they were about the size of the English siskin [about 4.75 inches long], and I judged the two to be male and female. One of them was quite dead, the other lay under the body of the spider not quite dead, and was smeared with the filthy liquor or saliva exuded by the monster. I drove away the spider and took the birds, but the second one soon died. The fact of [spiders] sallying forth at night, mounting trees, and sucking the eggs and young of humming-birds, has been recorded long ago by Madame Merian . . . but in the absence of any confirmation, it has come to be discredited.

Several South American spiders that are commonly known as tarantulas are large enough to capture birds, and it is likely that some of them do so from time to time, but it is not clear that Merian and Bates referred to the same species of spider.

The South American species that are called tarantulas are not to be confused with the much smaller tarantula of Europe, an unrelated and very different spider that derives its name from the city of Taranto in southern Italy. The bite of this spider was, quite falsely, believed to cause a disease known as tarantism. From the fifteenth to the seventeenth centuries, tarantism hysteria caused people who imagined that they had been bitten by a tarantula to break into a frenzied dance, the tarantella, that was thought to be both a symptom of and the cure for tarantism.

How large are bird-eating spiders? Bates wrote of an individual with a 7-inch leg span, but according to the 1995 *Guinness Book of World Records,* South American tarantulas grow to be much larger. Reputable collectors report that the largest known spider is a male goliath bird-eating spider *(Theraphosa leblondi)* that was captured in a coastal rain forest of northeastern South America. This individual had a leg span of 11.2 inches; he was large enough to cover a dinner plate. A female of the same species is the heaviest known spider. Her leg span was only 10.5 inches, but she weighed a whopping 4.3 ounces, considerably more than many small birds. By contrast, a ruby-throated hummingbird weighs only about 0.14 ounce, an average wood warbler about 0.35 ounce, a scarlet tanager about 1.0 ounce, and even a blue jay only about 3.5 ounces.

# The Birds Fight Back

6

For 15 years Mary Groff and Hervey Brackbill watched the peculiar behavior of common grackles that gathered in large and noisy flocks on English walnut trees near Lancaster, Pennsylvania. The birds visited these trees from early June, when the nuts were about three-quarters grown, until the middle of August. They did not come to feed. As Groff and Brackbill reported, the grackles would alight on clusters of nuts, one bird to each cluster, "and begin pecking a hole in the sticky hull of one of the nuts, usually throwing away the pieces of hull they gouged out but occasionally seeming to swallow a piece. When a good-sized hole had been made, the birds would dip their bills into it, undoubtedly wetting them against the pulpy interior, and then thrust their bills over and into their plumage. A great part of the body was thus anointed—the breast, the under and upper surfaces of the wings, the back, and very often apparently the rump at the base of the tail."

Common grackles are not the only birds that rub their plumage with foreign substances. Cardinals, blue-winged warblers, indigo buntings, American robins, orchard orioles and at least 200 other species of birds—almost all of them songbirds—are known to anoint their feathers with one or more of a long list of creatures or substances. The creatures include ants of at least 24 species, earwigs, grasshoppers, true bugs, wasps, and millipedes, all of which spray or exude pungent or repugnant chemicals—formic acid in the case of many of the ants. The substances—which all have strong, acrid, or repugnant odors—include, in addition to walnut husks, cigar and cigarette butts, burning matches, mothballs, oranges, lemons, limes, raw onions, marigold blossoms, soapsuds, vinegar, prepared mustard, beer, and hair tonic. Since birds most often use ants to anoint themselves, this behavior has come to be known as anting.

An anting bird may rub an ant on its plumage only once, or it may use the same ant repeatedly, rubbing it against only one wing or alternating from one wing to the other. Some birds, including some crows and starlings, use a wad of several ants. An ant may or may not be crushed, and after it has been used, it may be either discarded or eaten.

Songbirds ant in two stereotypical ways that differ little from species to species. A bird using the *active* method of anting stands near or among a group of ants and, as K. E. L. Simmons wrote, picks up one or more worker ants and "applies them with a quivering action of the bill to the undersurface of one wing at a time," particularly to the tip of the wing. The anting bird assumes a special and typical posture, one wing extended, its upper arm raised, the elbow "sprung forward and the hand partly extended to sweep the primaries [flight feathers] forward with their inner surface roughly parallel with the side of the body, and the spread tail thrust sideways and forwards to press up behind the raised wing."

A few birds, mainly some crows, thrushes, and finches, practice *passive* anting. The bird squats or lies among the ants, usually on top of an ant nest, and assumes a posture that maximally exposes its plumage. Then the bird either wallows with wings and tail spread-eagled or, as Simmons describes it, strikes a "more upright stance with both wings thrust forward in front of the body so that their inner surfaces are facing." The bird shakes its wings and tail and frequently repositions its body, movements that seem designed to stimulate the ants to greater activity.

In a note in *British Birds,* W. Condry described the passive anting behavior of a hand-reared carrion crow near a nest of ants that he had exposed by moving a slab of stone, "the crow displayed an immediate pronounced excitement. After a few seconds hesitation, he stepped into the middle of the swarming ants . . . When some of the ants found their way via his legs to his feathers, the bird showed apparent pleasure and slowly settled down among the ants like a brooding hen, with wings outspread and tail fanned. Then he dropped his head down in a swooning posture till his beak touched the ground."

Why do birds perform the complex, stereotypical, and apparently instinctive behaviors of active or passive anting? Several explanations have been suggested, but to my mind the most reasonable one is that anting helps the bird rid itself of external parasites. Lessening the depredations of parasites will, of course, improve the bird's chances for survival and reproductive success, and a behavior that serves this end will be favored by natural selection. I will come back to this explanation later, but first let us

A blissful crow spreading itself out on an ant nest

consider some of the alternative—often highly improbable—explanations. One is that the bird stores ants in its plumage so that it can eat them later. Another is that it rubs an ant against its feathers to wipe away distasteful substances on the ant before eating it. Neither of these explanations is supported by evidence, and there are good arguments against both of them. After dismissing other explanations, Lovie Whitaker, writing in *The Wilson Bulletin,* an ornithological journal, proposed another explanation, a farfetched one, involving the "pleasure principle." Arguing that birds most often apply ants to the area of the urogenital orifice, and that ant secretions have "thermogenic properties," she suggested that anting is an auto-erotic behavior. "Ant secretions might cause a peculiarly pleasurable sensation of warmth, possibly with an element of the masturbatory in it." But, as Simmons pointed out, it is highly improbable that natural selection would favor a complex instinctive behavior that gives pleasure but does not enhance an individuals probability of surviving to reproduce.

The objective of an anting bird is not the ants themselves, but the toxic chemicals that they produce. Birds ant only with those species of ants that secrete formic acid or other toxic fluids, and they are stimulated to ant by the odor of these substances. A European jay squirted in the face with formic acid by a puss caterpillar, wrote Niko Tinbergen, assumed the stereotypical posture of an anting bird. Simmons reported that "the manner of applying the ant is most sophisticated and assures the distribution of the toxic fluid on the feathers. The bird holds the ant by the thorax, leaving free the gaster, which contains the formic-acid apparatus and is roughly equivalent to the abdomen. The bird turns and applies the ant's gaster to the appropriate wing. It first holds the still live and struggling insect as it reflexively sprays formic acid onto the ventral surface of the bird's primaries, but the ant is eventually crushed in the bill by repeated nibbling and its oozing fluids are applied again and again, often on each wing alternately."

There is no doubt that formic acid is toxic to bird lice and mites. A German experimenter who sprayed living lice with formic acid found that they died within a few minutes, whereas control lice, sprayed only with water, were not affected. And as Leon Kelso and Margaret Nice have reported, Vsevolod Dubinin, a Russian entomologist, made some revealing observations of mites on meadow pipits. In the summer of 1943, Dubinin watched four of these birds on ant hills picking up wood ants and smearing them on their wing feathers. After 20 to 40 minutes he collected them and four other meadow pipits that had not been anting. Upon examining the birds that had been anting, he found that they smelled of formic acid, noted drops of what was presumably formic acid on their feathers, and found 90 dead feather mites in the vicinity of these drops. The rest of the mites on these four pipits were disturbed and were wandering over the feathers at random. The mites on the four pipits that had not been anting were undisturbed, resting in rows in the spaces between the barbs of the feathers. No dead ones were found. Dubinin noted that "of 642 live feather mites taken from the four anting pipits, 163 died within 12 hours, and eight more within 24 hours." But of 758 feather mites taken from the pipits that had not anted, only 5 died within 12 hours and only 2 more within 24 hours.

In July of 1990, Dale Clayton of the University of Utah watched a common grackle anting with a piece of lime fruit in Illinois, prompting him to test the effect of limes on bird lice. Of 52 lice on a feather placed in a covered glass petri dish with a slice of lime, 35 had died 9 hours later and the rest appeared to be on the verge of death. But, after 9 hours, only one out of 31 lice held in a control dish with moist tissue paper had died and all but one of the remaining lice appeared to be healthy. Clayton went on show that the toxic substance, whatever it is, is in the lime peel. Lice brushed with juice from the lime pulp were not effected, but all the lice brushed with an extract of the peel died within seconds.

✱ Birds have other grooming behaviors that help keep their plumage in good condition by cleansing it and ridding it of parasites: bathing, dusting, preening, oiling, scratching, and sunning.

With the probable exception of some desert species, it is likely that all birds bathe in water. There are several different methods of bathing, at least seven, all nicely described by Bruce Campbell and Elizabeth Lack in *A Dictionary of Birds*. Most birds practice "stand-in bathing." As it stands in the water, the bird ruffles its feathers and repeatedly dips its head and breast into the water as it shakes from side to side and flicks both wings up and forward. Then it wallows in the water with head raised, hind end submerged, and tail fanned as it flips water onto its back with one wing at a time. Other birds, kingfishers and certain tyrant flycatchers, plunge into the water from a perch. Swifts and swallows bathe in flight by dipping into the water. Some birds bathe in the rain, and others rub themselves against wet foliage. Most birds use only one of these methods, but a few may use two or more.

A bird may take several baths each day, a few brief ones that probably serve only to cool the body or dampen the feathers to facilitate the preening and oiling that inevitably follow each bath, and usually one prolonged and thorough bath that cleanses the plumage by washing away foreign matter, including mites and lice. As Campbell and Lack say about songbirds, a bird may "turn itself into a mass of dishevelled, watersoaked feathers, repeatedly exposing its feather tracts, forcing water into the [space between tracts] and squeezing it through the tracts, thus rinsing the skin and feather bases most efficiently."

"Bobwhites delight in the dust 'bath,' and it is likewise a delight to see them so obviously enjoying themselves, for every move denotes satisfaction as they wallow about in the loose dust, keeping a cloud in the air above their heads while they shake and work about to make the soft dust trickle down through the feathers and over the skin." So wrote Herbert Stoddard in his comprehensive 1921 monograph, *The Bobwhite Quail.*

Do not jump to the conclusion that these quail take dust baths only for the fun of it. I do not doubt that birds enjoy dust baths, anting, or other forms of grooming. They may even, as Lovie Whitaker proposed, derive sexual pleasure from anting. But it is most unlikely that the attainment of pleasure is the ultimate purpose, in an evolutionary sense, of these and other animal behaviors. Grooming behaviors are complex, stereotypical, and fixed in the genes. How could they have been favored by natural selection if they do not have some adaptive value—if they do not aid the bird in its quest to reproduce itself by helping it to survive and eat and grow? I believe that a feeling of pleasure is the way in which the brain rewards the body of a human, bird, or any other creature for performing an adaptive and necessary behavior that has no immediate reward. For example, nature induces us to reproduce ourselves—a reward that is delayed for nine months—by making sex so pleasurable.

Many but not all birds take dust baths. This behavior is particularly prevalent among birds that live in dry areas—deserts, savannas, or the like—where water for bathing is scarce; but many that live where water is readily available also dust themselves. Among the birds that take dust baths are the ostrich, rheas, buttonquails, bustards, seedsnipes, sandgrouse, doves, hawks, owls, nightjars, motmots, bee-eaters, rollers, hornbills, and some songbirds, such as larks, wrens, the common grackle, and some sparrows, such as the vesper sparrow.

Birds use several different methods of dusting, suggesting that this behavior evolved independently several times. House sparrows use their legs to loosen the earth and form a "dusting hollow," while quail, grouse, and pheasants use their bills to accomplish the same end. House sparrows use their wings to flick dust onto the plumage; hoopoes use mainly their feet; and white-winged choughs, Australian birds not related to the choughs of Europe, use their bills.

There is no doubt that dusting and the other grooming behaviors keep a bird's plumage in good condition. When a bird shakes the dust from its

body, it rids its feathers of the excess oils, mostly oil from the preen gland, that have been absorbed by the dust. Indeed, if birds are deprived of opportunities to take dust baths, their plumage becomes excessively oily and matted after only a few days. The little evidence that is available indicates that dusting also rids the plumage of external parasites such as lice. Miriam Rothschild and Theresa Clay reported that lice were found in the dust from dust baths that were used by chickens. Noting that bobwhites usually dust themselves from one to several times a day, Stoddard reported that captive bobwhites become very lousy if they are confined for two or three weeks without dusting facilities, and that wild individuals have more lice in wet weather, when they cannot dust, than during dry periods, when they can dust freely.

All birds preen, manipulate their feathers with the bill, especially immediately after a bath. Preening removes debris, rearranges the feathers, and anoints them with oil from the preen gland. There is also convincing evidence that preening, and head scratching as well, kill external parasites and thus help to keep their populations at low levels.

That preening birds attack the lice that infest their bodies is strongly suggested by adaptations of the lice that probably function to counter unwelcome attention from their host. Most lice are greatly flattened so that they can press themselves close against a feather or the skin to avoid the preening bill. Others are slender and elongate so that they can hunker down out of the way between two barbs of a feather. Some lice match the color of the feathers on which they live, probably to make themselves less visible to the host. Miriam Rothschild and Theresa Clay pointed out that a yellow louse lives on the yellow feathers of the golden oriole, a black one on the black feathers of the coot, and that there are white lice on white swans and black ones on the black swan and white lice on white gulls but dark ones on the related but dark colored "skuas" (some of which are known as jaegers in North America).

There is also direct proof that preening keeps populations of external parasites in check. Herbert Stoddard found feather lice in the stomachs of bobwhites, and several observers have reported extremely heavy infestations of lice on birds that could not preen properly because they had deformed bills. For example, Dale Clayton reported that a house sparrow that was missing the upper mandible of its bill was infested by over 1,200 lice and over 400 mites. In contrast, normal house sparrows collected in the same area hosted a mean of only 20 lice and 38 mites. Similarly, birds with deformed or missing feet often have unusually large numbers of lice on their heads and necks, areas that can be groomed only with the feet.

Clayton's recent experiments with rock doves (feral pigeons) leave no doubt that preening decreases louse populations. He inserted small metal "bits," lightly crimped to the nostrils, between the two mandibles of the bill, thus making it impossible for the bill to close all the way. These bits had no effect on the weight gain or the reproductive success of louse-free pigeons, but they did prevent them from preening normally. Clayton then compared the louse populations on birds with bits with those on control birds that had not been fitted with bits. In one experiment, the birds were held in cages, but in another they were released into the wild population of rock doves. After 15 weeks some captive and some free-ranging birds were killed and censused for lice. The results were striking. In both the captive and free-ranging groups, lice were significantly more numerous on "bitted" birds than on unimpaired birds. Free-ranging unimpaired birds bore a mean of 148 lice of one species and a mean of 133 of another species, but bitted birds bore a mean of 1,847 lice of the first species and 975 of the second species.

Clayton went on to demonstrate one of the ways in which reducing the population of lice can benefit a bird. He showed that the two species of lice on his rock doves, both of which feed on the abdominal contour feathers, destroy much of the plumage. There was a significant correlation between the number of lice on a bird and the total weight of its feathers, the higher the population of lice, the lower the total weight of the plumage. In one experiment, the presence of a large population of lice reduced the total weight of the plumage by 23 percent, as compared with birds that were free of lice. In a more recent similar experiment, the reduction in feather weight was 28 percent. Such reductions in the quantity of plumage must surely interfere with the bird's ability to control its body temperature.

Birds spread their wings and sun themselves on cold days, a behavior that certainly gives them much-needed warmth when temperatures are low. But as Tom Cade has pointed out, they have also been seen to sun themselves on warm days—even on hot days when the thermometer stood at 95°F, and they had to pant to keep from getting overheated. Consequently, other

functions have been proposed for sun bathing. In the case of griffin vultures, for example, it is probable that flight feathers deformed by the stresses of flight can be restored to their original shape by exposure to the sun. Another possible function, suggested by the fact that preening always follows sunning, is that the heating of the plumage makes lice and other external parasites uncomfortable, causing them to move and thus making them more accessible to the bill of a preening bird.

There is an intricate and mutually beneficial relationship between the bull's-horn acacia trees of the New World tropics and the stinging ants that inhabit them. The plant provides the ants with the hugely swollen thorns that they hollow out for nests and with nectar and special morsels of protein food that grow as appendages of the leaves. Colonies of these ants seldom survive away from an acacia, and the plant eventually dies if it is not occupied by a colony. The fiercely stinging ants reciprocate by killing leaf-feeding insects, by driving away browsing vertebrates as large as deer and cattle, and by pruning away invading plants that would otherwise compete with the shrub that houses them.

In 1945, the ornithologist Alexander Skutch observed that "birds of . . . numerous kinds place their nests in bull-horn acacias" and speculated that they are "taking advantage of the protection afforded by the stinging ants that dwell in the thorns." Some years later, Daniel Janzen further explored the ant-acacia relationship by counting the bird nests seen in acacias and other trees and shrubs that grew along the sides of more than 22 miles of a road in southern Mexico. He and another observer spotted nests on both sides of the road from a moving car. Although the car was going 20 to 25 miles an hour, they probably made a tolerably accurate count because it was still the dry season and nests were easily visible because the trees and shrubs had lost most of their leaves.

Janzen found 253 nests, 168 hanging nests built by orioles and caciques and 85 nests not of the hanging type that had most likely been built by several different species of birds, probably including wrens. Of these 253 nests, 159 were in bull's-horn acacias, a truly astonishing concentration when one considers that acacias were relatively scarce along the roadside, constituting, by Janzen's estimate, only one tenth of 1 percent of the vol-

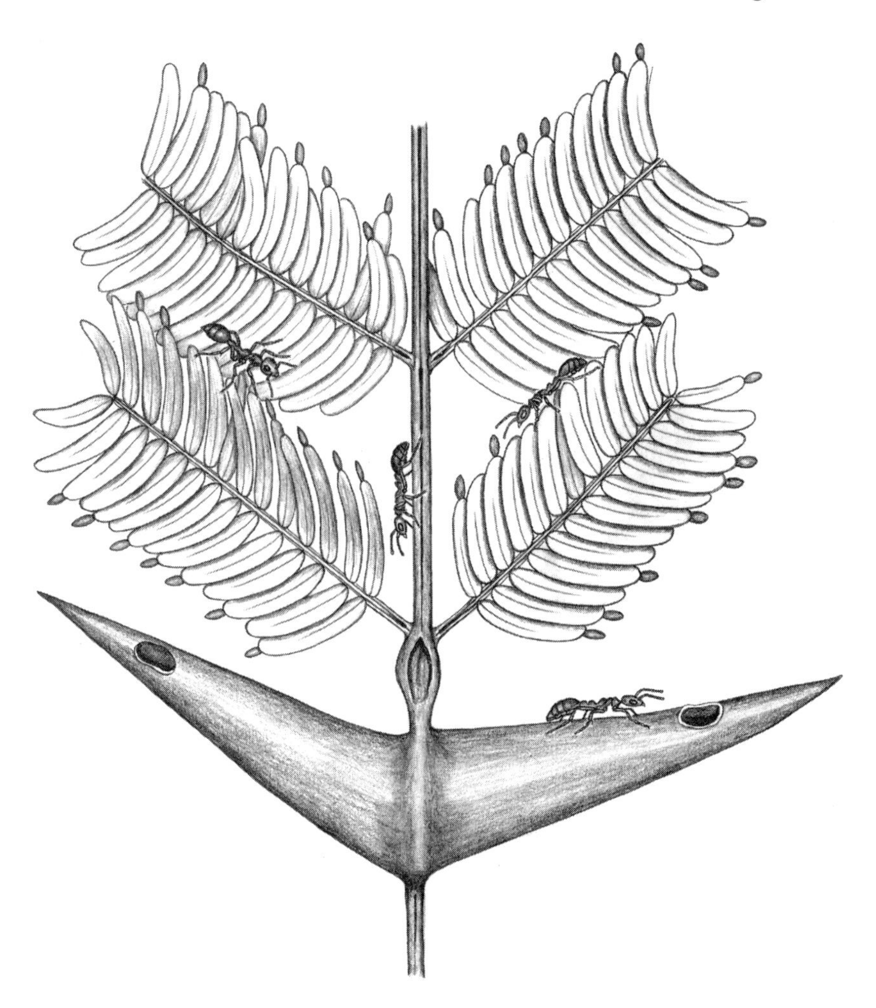

Stinging ants nesting in the thorns of a bull's-horn acacia; note the protein bodies at the tips of the leaflets, the nectary on the stem just above the thorns, and the nest entrance on each thorn

ume of vegetation present. Thus less than one one-thousandth of the vegetation contained over 66 percent of the nests. Janzen concluded, quite correctly in my opinion, that birds seek out acacias as nest sites.

Why are nesting birds not driven away by the ants? The ants seldom sting birds that nest in their tree, probably because they become habituated to the way the birds move and to the relatively gentle disturbances that they cause as they go to and from their nest. But other birds or mammals that

invade the tree are likely to be stung and driven away because they move differently and cause more noticeable disturbances. When Janzen pulled a branch down slightly to look in a nest, the worker ants on that branch became very disturbed and rushed about, but he noted that there was no such disturbance when nesting birds landed lightly on a branch.

There is little doubt that birds prefer to nest in bull's-horn acacias, and it seems inescapable that they must get some benefit from doing so. Janzen agrees with Skutch's hypothesis that the birds benefit because the ants protect their eggs and nestlings by driving away potential mammalian and avian predators. I have little doubt that this hypotheses is correct, but there is no direct evidence to support it. But, as you will read below, there is convincing evidence that birds benefit from nesting near colonies of stinging wasps, because the wasps drive away vertebrate predators.

At least 100 species of birds, mostly songbirds of both the Old and New World tropics, often build their nests close to colonies of wasps, ants, or bees, usually within about 5 feet. Included among them are some cuckoos, tyrant flycatchers, Old World flycatchers, tropical ovenbirds, sunbirds, wrens, flowerpeckers, bananaquits, orioles, weaver birds, and finches. The wasps, ants, and bees that are chosen as companions by these birds are all social species that form large colonies, and they are generally fierce stingers that are quick to defend their nests against mammals or other vertebrates that come too close.

Some of these birds associate with colonies of stingless bees (meliponines), but don't be misled by their name. Although they have a vestigial stinger that cannot be used in defense, these bees are fully capable of providing a defensive shield against predators. As E. O. Wilson wrote in *The Insect Societies*, they painfully repel people who come too close to their nest, swarming "over the body, pinching the skin and pulling hair, occasionally locking their mandibles in catatonic spasms so that before the grip can be broken, their heads tear loose from the body." Some stingless bees of the genus *Trigona,* which are often sought out as nesting companions by birds in Central and South America, also eject from their mandibles a liquid so caustic that it can burn off patches of skin.

Evidence that wasps actually do give their bird associates protection against nest predators was published only rather recently. Surveys of naturally occurring bananaquit nests on the West Indian island of Grenada, done by Joseph Wunderle and Kenneth Pollock, showed that nests built close to wasp colonies that hung in the same tree were less likely to be raided by predators than were nests on neighboring trees that were not associated with a wasp colony. This observation and the similar observations from a few other studies of bird-wasp and bird-*Trigona* associations are quite compelling, but not as completely convincing as are the results of a controlled experiment done under natural conditions in Costa Rica by Frank Joyce.

Joyce found that rufous-naped wrens usually build their nests in bull's-horn acacias, sometimes close to a wasp's nest that hangs in the same tree. He then devised an experiment to find out whether nesting wrens associated with a wasp's nest were more likely to produce offspring that survived to leave the nest than were nesting wrens that were not associated with a wasp's nest. He first located a number of wrens' nests that had been built in acacias that did not contain a wasp's nest. One group of these nests, randomly selected, served as controls that were left as they were, not associated with a wasp's nest. The nests in another group, also randomly selected, were experimentally manipulated by hanging an active wasp's nest within about 2 feet of each one.

The results of this experiment leave no doubt that proximity to a wasp's nest dramatically increased nesting success over and above any protection afforded by acacia ants, almost certainly by giving protection against the white-faced monkeys that are the main predators of nesting rufous-naped wrens in the area where the experiment was done. (Working in Peru, Scott Robinson, an ornithologist employed by the Illinois Natural History Survey, watched as wasps repelled attacks on nests of yellow-rumped caciques by three different species of monkey.) In an experiment done by Joyce in 1987, no young wrens fledged from nests without a wasp colony, but over 37 percent of the nests with a wasp colony did produce fledglings. A second experiment, done the following year, gave corroborating results. Young birds fledged from only 20 percent of the nests without wasps, but fledglings emerged from 75 percent of the nests that had been provided with a wasp's nest. Joyce's experiments, as designed, cannot tell us whether or not the acacia ants gave the wrens any protection from predators. I think that they

probably did. However, it is obvious that the presence of wasps added to any protection that may have been given by the ants.

Stinging wasps, stingless but biting meliponine bees, flies that are parasites of birds, chestnut-headed oropendolas, yellow-rumped caciques, and giant cowbirds of the New World tropics are all involved in complex relationships with each other—some of which are mutually beneficial to the participants. These intricate interactions were discovered and unraveled by Neal Smith of the Smithsonian Tropical Research Institute in Panama, and nicely summarized by Robert Ricklefs in his textbook on ecology. The stinging wasps and the biting bees nest in trees, and their nests are often the focus of a mixed colony of anywhere from a few to a hundred or more oropendola and cacique nests that cluster closely around them, often within 10 feet or less. (Since, according to Smith, the oropendolas and caciques behave in much the same way, I will henceforth refer only to the oropendolas.) The parasitic flies, which weaken and sometimes kill oropendola nestlings, lay their eggs on the bodies of both nestling oropendolas and nestling cowbirds, and their larvae live as external parasites on the nestlings. Smith referred to these parasites as bot flies, but if he is correct in ascribing the generic name *Philornis* to them, they are actually relatives of the house fly, and are not related to the bot flies, which, as far as is known, are all internal parasites of mammals. The oropendolas are the hosts of the giant cowbirds, brood parasites that lay their eggs in oropendola nests and abandon their offspring to the care of the foster parents. Cowbird nestlings compete with their oropendola nestmates, and thus have a detrimental effect on them, although the oropendola nestlings usually do manage to survive.

After several years of intensive research, Smith had established that there are two quite distinct types of chestnut-headed oropendola colonies, whose members behave quite differently. In one type, the oropendolas nest as close as possible to colonies of wasps or bees, and try to avoid, often successfully, being parasitized by giant cowbirds by ejecting their eggs from the nest. Smith referred to these as discriminator colonies, but I think that it will help to understand this complex situation if we think of them as cowbird rejecters. When Smith experimented by putting model eggs and other foreign objects in the nests of cowbird-rejecting oropendolas, they removed all but those objects that were virtually indistinguishable from their own eggs.

In the other type of colony, the oropendolas do not nest in association with wasps or bees, and they do not eject cowbird eggs from their nests. Smith referred to such colonies as nondiscriminators, but I will call them cowbird acceptors. When he put model eggs or other foreign objects in the nests of cowbird-accepting oropendolas, they did not remove them—not even if they bore no resemblance to their own eggs.

The giant cowbirds have adapted to the two different types of oropendola colonies. Cowbirds that associate with cowbird rejecters are sneaky when they enter a colony, behaving timidly and unobtrusively. They also lay eggs that closely mimic eggs of the chestnut-headed oropendola, and lay only one egg per nest. Cowbirds that associate with cowbird acceptors are not sneaky. Quite to the contrary, they behave aggressively and sometimes enter oropendola colonies in groups. They do not lay mimetic eggs and may place several eggs in the same nest.

The differences between the two types of oropendola colonies arise because both the cowbirds and the wasps and bees are enemies of the parasitic flies. A cowbird nestling grooms its nestmates, be they oropendolas or fellow cowbirds, thus destroying the eggs and larvae of the parasitic fly. Wasps, which are predators of other insects, and even bees, which are not predators, deter adult flies from entering the oropendola colony. Exactly how they do this is not clear, but there is no doubt that they do, since severed wings of these flies can often be found on the ground beneath a wasp or bee colony.

Oropendola colonies that are not protected from the parasitic flies by wasps or bees benefit by accepting cowbirds, because the cowbird nestlings remove parasitic fly eggs and larvae from the bodies of the oropendola nestlings. In this case, the harm done by the cowbird parasite is outweighed by the greater benefit that it confers and natural selection favors the acceptance of the cowbirds. But oropendolas that nest near wasp or bee colonies are seldom troubled by the parasitic flies. If there are no parasitic maggots in the nest, a cowbird chick cannot mitigate the harm that it does by remov-

ing parasites from the oropendola chick. Thus when oropendolas nest near wasps or bees, natural selection favors those that reject cowbirds.

In April of 1976, S. Sengupta of the Zoological Survey of India watched a pair of house sparrows in Sinthee, a suburb of Calcutta, enter their nest with green leaves of the neem (margosa) tree in their bills. A search in the vicinity of this nest turned up another 25 accessible house sparrow nests, all of which contained neem leaves. The next year, Sengupta inspected nearly all of the accessible house sparrow nests in Sinthee, 120 of them, and found neem leaves in every one. But of the 230 accessible house sparrow nests that he found in a random search of nearby Calcutta, only 20 had neem leaves although neem trees were usually nearby. It seems that the local population of house sparrows in Sinthee comprised a "subculture" that had adopted the habit of putting neem leaves in their nests, but that this habit had not yet spread throughout Calcutta.

As you will soon read in more detail, many birds add fresh green plant material to their nests, especially leaves of plants that have insecticidal properties. Several investigators have proposed that this is a beneficial behavior that serves to kill or repel arthropod parasites of the birds. The neem tree is among the most potent of the insecticidal plants. As Ramesh Saxena of the International Rice Research Institute (IRRI) has written, the insecticidal properties of neem have been known in India since antiquity, and to this day neem seeds, leaves, and extracts are used to control many different kinds of insects. Neem leaves pressed in books are said to keep out booklice and other insects. In Indian households, neem leaves are used to protect woolen clothing and stored grain from being attacked by beetles and caterpillars. Extracts of neem are sprayed on cotton and other crops to prevent damage by whiteflies, cutworms, and various other leaf-feeding caterpillars.

During my 1978-79 sabbatical year at IRRI, near Los Baños in the Philippines—home of the "green revolution" in rice farming—I had first-hand experience with the effects of neem on insects when I took time from my main research project to join Saxena and his research team in exploring the use of neem seed oil to control the rice leaf folder, a pest of the rice crop throughout much of Asia. Spraying the oil on rice leaves killed eggs and either killed or repelled caterpillars and adult moths. Caterpillars exposed to neem oil did not survive the larval-pupal metamorphosis, some of them molting to become monstrous intergrades, destined for death, between the larval and pupal stages rather than normal pupae. This suggests that neem oil contains a hormonelike substance that disrupts the hormonal system that controls metamorphosis.

Not only house sparrows, but also some ducks, hawks, anis, crows, purple martins, starlings, and quite a few other birds add fresh green vegetation of various kinds to their nests. The hypothesis that this behavior has been favored by natural selection because it reduces parasite populations is supported by the fact that birds that reuse their nests or nest in cavities are especially likely to add green vegetation to their nests. This may well be because they are likely to be plagued by larger populations of parasitic mites and fleas than are other birds, because these parasites can survive for a long time in exposed unused nests or in cavities, even through the winter. An analysis by Peter Wimberger revealed that, of 49 eagles, kites, and hawks, 30 species reuse their nests year after year and 19 do not. Of the nest reusers, 22 (73 percent) put green plant material in their nests, but only 6 (31 percent) of the 19 species that build a fresh nest each year do so. Larry Clark and J. Russell Mason did a similar analysis of 137 species of songbirds that nest in the eastern United States and Canada. Of these birds, 27 species nest in enclosed spaces and 18 (67 percent) of them add green vegetation to their nests. But of the 110 species that build open, cuplike nests, only 28 (25 percent) add green vegetation to their nests.

A study of cavity-nesting European starlings by Clark and Mason showed that the green plants that they added to their nests were not chosen at random, but instead constituted only a small subset of the many plants that were available near the nest. The plants selected were generally those that contain volatile biochemicals that slow down or stop the growth of bacteria and kill arthropod parasites or otherwise reduce their populations. Among these plants were wild carrot (also known as Queen Anne's lace) and fleabane, both known to be insecticidal, and the latter bearing a traditional common name that bespeaks its insecticidal value. Clark and Mason showed that starlings have a much better sense of smell than was once thought, that they can be conditioned to respond to volatile plant compounds, and that they almost certainly use their sense of smell to choose the foliage that they put in their nests.

In an another experiment, also done under natural conditions, Larry Clark demonstrated that fresh green foliage can greatly decrease the population of tiny blood-sucking mites in starling nests. He added green wild carrot foliage to one group of nests every day. And from another group he removed all the fresh foliage every day. During the first breeding attempt in May, nests protected by wild carrot foliage harbored an average of 3,000 mites, but nests not so protected had an average of 80,000 mites. By the second breeding attempt in June, mites had become far more numerous in these nests, but wild carrot foliage still gave considerable protection. Nests with fresh foliage contained an average of 11,000 mites, but nests without fresh foliage contained an astonishing average of 500,000 mites.

The deleterious effect of these blood-sucking mites on starlings is subtle and may not be apparent until long after the young birds leave the nest. It is therefore difficult to demonstrate the benefits gained from reducing the population of mites. Mortality, weight gain, and fat content of newly fledged chicks did not differ, but starling chicks from nests not protected by wild carrot foliage were anemic compared with chicks from nests that were protected by green wild carrot foliage.

# Bugs That Eat People

# 7

On the 1960 trip to the Texas coast that I mentioned earlier, I was accompanied by William Downes, a fellow entomologist but definitely not a birder. Bill, a fly specialist, was collecting flies of several families, and I, pursuing my interest in mimicry, was collecting flies of the family Syrphidae, the flower flies, sometimes called hover flies, which are often colorful and are sometimes spectacularly deceptive mimics of wasps and bumble bees. I birded in spare moments. One afternoon we pitched our tent on the deserted Gulf beach of the Bolivar Peninsula, only a few miles northeast of Port Bolivar. As Bill captured flies by sweeping the beach grass with his insect net, I sat on the sand with my binoculars, identifying the birds that flew or waded nearby—among them two lifers, several gull-billed terns, and one reddish egret.

Later we made a fire of driftwood, heated our meal of canned food, and watched the sun go down over the wide marsh on the landward side of the peninsula. As it began to darken, we noticed what we at first took to be a large cloud of smoke rising from the marsh. But we soon realized what we were really seeing: bugs that eat birders and entomologists!—a vast swarm of countless millions of salt marsh mosquitoes departing their larval habitat in search of blood meals. The breeze was blowing the mosquitoes toward us, so we immediately smeared ourselves with repellent. They soon arrived and proceeded to plague us almost beyond bearing. Hundreds of them hovered just an inch or so in front of our eyes. Some actually bit us despite the repellent, and when we later removed our boots before taking refuge in our screened tent, scores of them instantly landed on our repellent-free ankles and bit us mercilessly. Bats and common nighthawks flew through the light of our fire as they caught mosquitoes. Although it was almost dark, some day-flying predators, notably swallows and dragonflies, continued

their feeding frenzy in the glow of the fire. So numerous were these insectivores that the rustle of their wings, the snapping of their beaks, and the grinding of their mandibles were easily audible above the incessant, high-pitched hum of the mosquitoes.

It usually is not that bad, and sometimes birders are not troubled by insects at all. But as all birders eventually learn, almost any place can, in the right season and at the right time of day, have its share of insects or other arthropods that are annoying or even dangerous to human health. Chiggers literally eat our skin. Ticks, mosquitoes, black flies, and deer flies puncture us to get at our blood. A few true bugs do the same, and in southeast Asia there is even a very atypical moth that sucks the blood of mammals, probably including humans. All these bloodsuckers are annoying, and some of them can infect us with bacteria or other disease-causing organisms.

Insects and other arthropods that do not consume our substance may be a nuisance or even seriously threaten us in other ways. Some gnats can be so abundant that they cause severe discomfort by their very presence even though they do not bite, flying into our mouths or being sucked into our nostrils. Some noninsectan arthropods, including scorpions and certain spiders, inject venom when they sting or bite. Wasps, ants, and bees, especially social species such as hornets, yellowjackets, honey bees, bumble bees, and some of the ants, readily sting in defense of their colonies; but fortunately they rarely sting when they are away from the colony, where we most often encounter them.

* Thumb through the sales catalogue of the American Birding Association, and you will find advertisements for regional field guides to birds—often quite recently published and usually in English—that reflect the interest of many modern birders, especially North Americans and Europeans, in seeking out the birds of far-flung countries such as Nepal, Colombia, and Australia. Traveling birders and other people may be exposed, especially in the tropics, to annoying or even dangerous insects that are unknown in North America.

Tsetses, blood-sucking flies that attack humans, zebras, wildebeests, and other large animals of the savannas, are among the troublesome insects that people are likely to encounter in Africa. In infested areas people are so pestered by these flies that they carry whisks, which they keep in almost constant motion to brush the flies away from their bodies. The bite of a tsetse is not only painful and annoying. It may also transmit the protozoan, a trypanosome, that causes African sleeping sickness, a debilitating and sometimes deadly disease of humans.

The bush fly of Australia is closely related to the cosmopolitan house fly, that familiar but unwelcome visitor to our homes. Like house flies, bush flies have mouthparts that are adapted for sponging up thin films of liquid, but unlike those of the house fly, the bush fly's mouthparts are also equipped with sharp teeth that can rasp soft tissues so as to cause bleeding. During the daylight hours, bush flies make life miserable for people all over Australia, Tasmania, and the southern areas of Papua. In the warm months they are usually distressingly abundant, resting on any part of the body that is out of the wind and busily exploring the eyes, mouth, or wounds as they search for saliva, tears, blood, or pus.

As a larva, or maggot, the human bot fly of Central and South America, also known as the tórsalo, is a parasite that lives just beneath the skin in the flesh of humans, domestic and wild animals, and sometimes even birds such as toucans and turkeys. The lesion caused by the full-grown larva looks like a large boil, oozing pus and with the spiracles at the hind end of the maggot almost protruding from a hole in the skin. By contrast, the adult fly, which is rarely seen, is harmless.

Although other species of bots glue their eggs to the hairs of the host, the tórsalo, the only bot that regularly parasitizes people, uses an indirect method to get its eggs to a host. A gravid female captures a mosquito or some other blood-sucking arthropod and glues several eggs to its body. When this booby-trapped blood sucker later lands on a warm-blooded animal, the little tórsalo larvae, stimulated by the sudden increase in temperature, instantly pop out of the eggshell and quickly burrow into the skin of the host, where they will remain for about two months as they feed and grow to full size, about an inch in length.

My friend Carl Bouton, an entomologist and ecologist who spends much of his time doing field work in Peru, told me about a method, recommended by local residents, that he used to remove three small tórsalo larvae from the flesh of his arm. The area around the lesion is shaved, and a piece of adhesive tape is firmly applied to seal shut the small hole in the skin

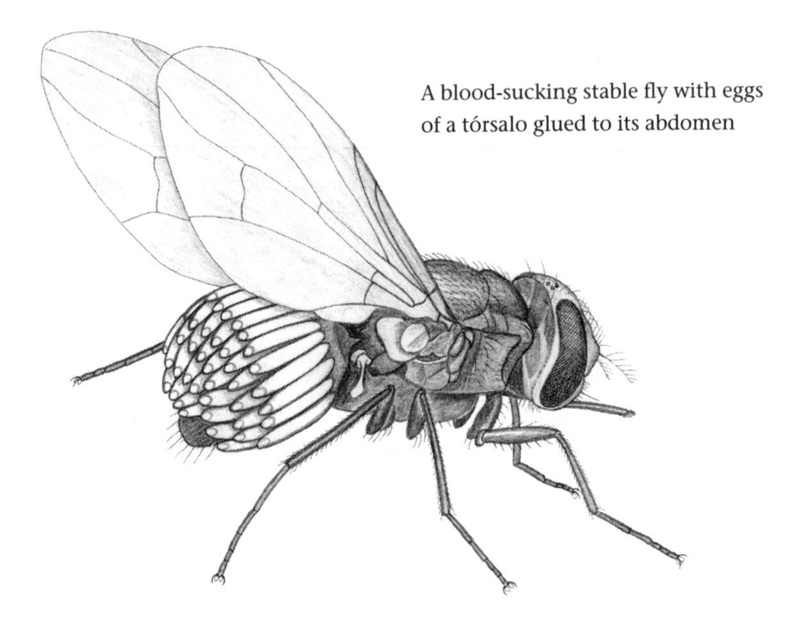

A blood-sucking stable fly with eggs of a tórsalo glued to its abdomen

through which the larva breathes. The suffocating larva, in search of air, crawls out of the skin, sticks to the adhesive, and is removed by gently peeling off the tape. I have heard of another method, one that I think would work well with large larvae. A thick piece of raw bacon is tightly bound over the lesion. The tórsalo larva, trying to reach air, burrows into the bacon, which can then be discarded. (I imagine bacon is used because it is slower to rot than would be a piece of fresh meat.)

Along swift-flowing streams in Africa, a species of black fly, aptly named *Simulium damnosum,* inflicts painful bites on people during the daylight hours. These flies may be so horribly abundant that they cause the death of livestock as a consequence of toxemia, blood loss, and suffocation caused by inhalation of the flies. Their bites and those of related species that occur in parts of Mexico and Central and South America transmit the tiny larvae of a nematode worm, *Onchocerca volvulus,* that cause the disease known as onchocerciasis, or river blindness. Adult worms and larvae occur in large, nodular tumors that are found on the head, trunk, and shoulders of infested people. A full-grown female worm may be as much as 27 inches long, but the males are no more than 3.5 inches long. The minute larvae, which often migrate through the body, can cause blindness if they reach the eyes.

In the forests of South America you may encounter a huge, stinging ant, well over an inch long, that goes by the scientific name of *Dinoponera giganteus,* which translates from the Greek as "the gigantic, terrible pain." Its sting is said to be very severe, but the sting of a related ant of Central and South America, the tucandeira *(Paraponera clavata)* is said to be far worse, excruciatingly painful and even a threat to life. My friend and fellow entomologist John Bouseman told me of seeing the Uruguayan ornithologist Juan Cuello, coauthor of the 1962 *Las aves del Uruguay,* being painfully stung by tucandeiras on the wild banks of the Rio Machupa in Bolivia. Cuello was stung by several of these ants when he climbed a tree to examine some oropendola nests. As he rushed down the tree he tore off his clothes, and then jumped directly into the river to rid himself of the ants. (I was tempted to write *piranha-infested* river, but that would have been too sensational. It is safe to assume that there were piranhas in the river, but people often swim in piranha-infested waters without being attacked.)

Before discussing the biting and stinging pests that you are likely to encounter in North America, I will tell you about an insect that occurs in California and Oregon, a gnat, that neither bites nor stings but that can be unbearably annoying when it occurs in large numbers, as it sometimes does under certain local conditions. The attempts to control this insect by chemical means constitute a grim but informative chapter in the long history of the ecological disasters brought on by the indiscriminate use of pesticides by the ecologically uninformed.

Clear Lake, in the mountains of northern California, is a large, shallow, and—belying its name—somewhat turbid body of water with a muddy bottom. Conditions in this lake often permitted the production of hundreds of millions of its namesake insect, the Clear Lake gnat. Although it does not bite, this gnat was sometimes so abundant that its very presence made life miserable for anglers and other holiday visitors to the lake. The lake supported game fish, such as largemouth bass, and over 1,000 pairs of fish-eating western grebes that built floating nests among the reeds at its margin.

In order to control the population of gnats, DDD (also known as TDE), a nonbiodegradable chlorinated hydrocarbon insecticide closely related to DDT, was applied to the lake from 1949 to 1958. DDD did indeed kill most

of the gnats, although at the end there was good reason to suspect that they were becoming resistant to it—as many hundreds of insects have become resistant to virtually all of the many insecticides that have been devised. As Rachel Carson wrote, the assault on the Clear Lake gnat was well planned, and only a few people could have predicted the disastrous consequences. Among these few people were ecologists who would have been aware of the nature of food chains. But no ecologists were consulted.

The lake was measured, its volume calculated, and DDD applied at the maximum rate of only 0.02 of a part per million—just one part of DDD in 50 million parts of water—a concentration sufficient to kill the larvae of Clear Lake gnats but far too low to directly harm fish, birds, or other vertebrates. Nevertheless, some time after the application, many bass died, and hundreds of dead western grebes were found along the shore. By 1958 only about 30 pairs of these birds survived on the lake. They tried to breed, but their hatchlings died. The grebes were found to be free of disease and their deaths remained a mystery until it finally occurred to someone to examine their bodies for DDD. The results were absolutely astonishing.

The body fat of the grebes contained an average residue of well over 2,000 parts per million of DDD, over 100,000 times its concentration in the water of the lake. How did this happen? The answer is that the essentially nonbiodegradable and persistent DDD, which is stored in the body fat of animals, was concentrated by the food chain that supports the grebes. The plankton, which consist mainly of unicellular, floating green algae in Clear Lake, are at the beginning of this food chain. They absorbed the insecticide from the water and concentrated it to about 10 parts per million in the fat of their bodies; herbivorous fish ate plankton and accumulated about 900 parts per million of DDD in their body fat; carnivorous fish, such as the bass, ate the herbivorous fish and, in turn, stored almost 2,700 parts per million in the fat of their bodies. The grebes, which eat a mix of herbivorous and carnivorous fish, ended up with about 2,130 parts per million in the fat of their bodies, enough to cause death.

In 1959, the California Department of Health banned the use of DDD in Clear Lake. At the same time, however, department officials claimed that DDD did not present a health hazard—despite the fact that people were eating Clear Lake bass containing enormous concentrations of this toxic chlorinated hydrocarbon. But now we know better. For example, states bordering the Great Lakes now publish warnings about eating Great Lakes salmon, which contains far lower but still worrisome concentrations of chlorinated hycrocarbons.

The last application of DDD to Clear Lake was made in 1958, but the poison lingered in the lake for years, although none could be found in the water. All of it was held in the bodies of the plants and animals that lived in the lake. In 1961, three years after the last application of DDD, only 16 nests of the western grebe were found on Clear Lake, and none of the young from these nests survived.

A more environmentally friendly way of reducing the population of the Clear Lake gnat was finally found, as Mary Louise Flint and Robert van den Bosch reported in their book on pest management. In the late 1970s, a small fish, called the inland silverside, that eats the larvae of the gnat was introduced into the lake. Since then the Clear Lake gnat has not been a problem, and the population of grebes has been on the increase. Insecticides failed, but a fish seems to have saved the day. A word of caution, however—introducing any foreign animal into an ecosystem may have detrimental and unforeseen consequences, as did the unfortunate introduction of the starling and the house sparrow to North America.

There are a few other instances of birds dying as a direct result of exposure to pesticides, but pesticide poisoning usually takes a more subtle and insidious form—the inhibition of reproduction. Inhibition may occur via different physiological routes, but the most apparent in birds is the laying of eggs that have abnormally thin shells because of the interference of chlorinated hydrocarbons, such as DDT, with the metabolism of calcium, the major constituent of the eggshell. When a parent bird tries to incubate a thin-shelled egg, the shell breaks and the egg is lost. Raptors and fish-eating birds, which are at the top of their food chains, are especially prone to eggshell thinning; among them are American kestrels, peregrines, golden eagles, brown pelicans, bald eagles, and ospreys. Populations of all these birds were severely depleted by chlorinated hydrocarbon insecticides such as DDT, dieldrin, and toxaphene.

The peregrine was actually extirpated from the eastern United States. But it has been reintroduced to the East and its population and the populations of the other birds mentioned have been recovering since the banning of DDT by the United States in 1972 and the subsequent banning of other

chlorinated hydrocarbon insecticides. Today, peregrines even live in cities, eating pigeons and nesting on buildings. In 1996 a friend and I sat in the lobby of an office building in Toronto, watching on closed-circuit television a pair of peregrines that were raising four young on a nineteenth-floor ledge of that same building.

Among the insects that actually bite, pierce, or otherwise attack humans are the biting midges, members of the Ceratopogonidae, a cosmopolitan family composed of tiny flies that, as you read in a previous chapter, are quite diverse in their feeding habits. Some of them eat mainly nectar and other plant exudates. But the females of quite a few species, most of them members of the genus *Culicoides,* suck the blood of humans. These blood suckers are variously known as punkies or no-see-ums—the latter an apt name reportedly derived from a Native American language.

Punkies are minute, generally ranging from about 0.04 to 0.08 inch long, small enough to pass easily through 16-mesh mosquito screening. Their wings, marked with gray or brown patches, are held down flat when they sit on our skin to suck our blood. A punkie on the back of your hand looks much like a tiny fleck of ash.

Richard H. Foote and Harry D. Pratt reported that these pestiferous flies bite mainly around sunset, from about 4:00 P.M. to about 9:00 P.M., but that they also bite through the daylight hours. The pain resulting from a punkie's bite is worse than what its small size might lead you to expect. Some people think that it is worse than the bite of a mosquito or a black fly. There is a sharp, burning pain, an area of redness surrounding the bite, and sometimes considerable swelling. Some people suffer severe reactions that may persist for several days.

Punkies, generally most prevalent from mid to late summer, occur in areas close to the habitat of the larvae, which grow in decaying vegetation and silt, in brackish streams, along the margins of ponds and streams, or in wet rot cavities in trees. In my experience, punkies are very often abundant in mountain areas. J. H. Comstock wrote that they are exceedingly troublesome in the Adirondacks in New York state, the mountains of New England, and along mountain streams generally. I remember being plagued by punkies in 1952 as fellow birder Anne Freeman and I watched a family of red-breasted mergansers swim on the Mad River in the White Mountains of New Hampshire.

Mosquitoes are the blood-sucking insects that we most often encounter in North America. Their bites are annoying and cause severe allergic reactions in a few people. Although the major mosquito-borne diseases of humans—malaria, dengue, filariasis, and yellow fever—are virtually absent from the United States and Canada, North American mosquitoes do transmit viruses that cause other significant, although not very prevalent, diseases: LaCrosse fever and eastern, western, California, and St. Louis encephalitis.

Adult mosquitoes are agile fliers that search for blood meals and mates on the wing. The larvae, often called wrigglers, and the pupae, known as tumblers, are aquatic, and both are active swimmers, an exception to the generalization that insect pupae are inactive. My friend and mentor, William R. Horsfall, the dean of North American medical entomologists, taught me most of what I know about mosquitoes and impressed upon me the fact that wherever there are quiet waters, there are likely to be mosquito larvae. The larvae avoid currents and wave action, but they can be found in the backwaters of lakes and ponds, and even in quiet eddies in streams, especially among the stems of cattails or other plants that grow in standing water. Bill Horsfall tells me that they occur even in icy pools in Canada only 7 degrees south of the North Pole. Mosquito larvae also live in the liquid in pitcher plants, in water-filled cavities in trees, in stagnant birdbaths, and in the water in discarded beer cans. And the Asian tiger mosquito, a recent and unwelcome introduction to North America, is likely to be among the mosquito larvae that live in the stagnant water in discarded tires. In fact, it made its way here in used tires imported from Asia.

Mosquitoes lay their eggs in the habitat of the larvae, often floating them on the surface of the water. *Anopheles* mosquitoes, the vectors of malaria, lay their eggs singly, each egg equipped with two flotation devices that prevent it from sinking. *Culex* females form floating rafts that consist of dozens of eggs that cling closely together. But many species of the genera *Psorophora* and *Aedes* are known as floodwater mosquitoes because they lay their eggs in crevices in dry soil that is subject to periodic flooding. The eggs

do not hatch until some time after they are covered with water, as when rain fills a temporary woodland pool or when a river comes out of its banks to inundate its floodplain. Among these floodwater mosquitoes are some of our most pestiferous species, as is clear from the scientific names bestowed on some of them: *Aedes vexans, Aedes excrucians,* and *Psorophora horrida.*

A few larval mosquitoes prey on other mosquito larvae, but the great majority are filter feeders that use their fanlike mouthparts to set up a current of water from which they strain unicellular algae and other bits of organic matter. With only a few exceptions, they must come to the surface to obtain oxygen, using a snorkel at the end of the abdomen to penetrate the surface film of the water. A few species always remain below the surface, using the snorkel, sharp and spikelike in their case, to penetrate air-filled cavities in the stems of aquatic plants.

Many of us have encountered a mating swarm of mosquitoes, thousands of the tiny insects forming a closely packed swarm that hangs dancing in the air without shifting its position. These swarms are harmless. They consist of males, which do not bite, each one facing into the wind and hovering so as to maintain his position over some visible marker below, perhaps a small puddle, a stone, or a trail sign. Females orient to the swarm, fly into it, and are almost immediately grabbed by a male, who takes his conquest out of the swarm and retreats with her to nearby vegetation, where they hide for the few moments that it takes him to inseminate her.

*Aedes vexans,* known as the inland floodwater mosquito, is widely distributed in North America, occurring in Alaska, much of Canada, and all 48 of the contiguous United States. In many areas it is the major pest mosquito. It becomes abundant after heavy rains, but is scarce or absent when the weather has been dry and most of its population consists of eggs in the soil. In a year when the pattern of rainfall is such that there are alternate periods of flooding and drying, several distinct waves—generations, broadly speaking—of these mosquitoes may appear during the warm months. Eggs are laid in the damp soil between floodings. Horsfall reports that in central Illinois there may be as many as three separate waves of ferociously biting adults during a single warm season.

The eggs of *Aedes vexans* and those of other floodwater mosquitoes will not hatch until they are covered with water, but immersion alone is not sufficient to stimulate the embryo to pop out of the egg. In 1934, an experimenter found that eggs held in a warm laboratory could be kept submerged in clean water for as long as 86 days without hatching. A few years later, other investigators found that floodwater mosquito eggs hatched if they were covered with water containing organic detritus rather than with pure water. It seemed likely that the growth of microorganisms in water containing organic debris stimulated the eggs to hatch. It turns out that this is indeed the case, a highly adaptive arrangement that assures that microorganisms are developing in the floodwater and will be available as food for the newly hatched larvae.

But what is it about the presence of bacteria and other microbes that signals the embryos to pop out of the egg? In 1953, Alfred E. Borg and Bill Horsfall showed that a decrease in the dissolved oxygen content of the water, brought on by the respiration of increasing numbers of microorganisms, stimulates hatching. They found that hatching could be triggered, even in the absence of microorganisms, just by lowering the dissolved oxygen content of the water, which they accomplished by removing the air over the water and replacing it with nitrogen. The dissolved oxygen from the water diffused into the nitrogen atmosphere and was thus not available to the eggs.

Not surprisingly, almost everyone focuses on the nuisance value of mosquitoes. I have even been asked what role mosquitoes could possibly have in the great scheme of things, the implication being that the world would be better off if all 2,500 species were to vanish forever. But as troublesome as they can be, mosquitoes play many significant ecological roles, and the world would probably be worse off without them. Let's look at just two of these roles. First, when an adult mosquito transmits a disease-causing virus, bacterium, or protozoan to a bird, opossum, rabbit, or some other vertebrate, it is helping to slow what might otherwise become runaway population growth. Second, larval mosquitoes are important links in many aquatic food chains. They filter microorganisms and tiny particles of organic matter from the water, incorporate the contained nutrients in their bodies, and thus grow to become packages of nutrients that are big enough to be eaten by fish and other small creatures that have no way to gather tiny microorganisms or minute particles.

But mosquitoes are also interesting in themselves—as exemplars of evolution, as "a sample of the living force that runs the earth," as E. O. Wilson

put it in *The Diversity of Life*. Once we get to know them as an entomologist does, we can appreciate how exquisitely mosquitoes are adapted to their way of life.

Consider the subtlety of the female mosquito's bite. More often than not we do not even notice her as she sucks blood from the back of our neck or some other uncovered or thinly covered area of the body. The pain and the itching come later, after she has finished feeding. Her usually painless bite is an important advantage to her, because a painful bite will elicit the lethal slap of a hand or the disturbing or sometimes even deadly swish of a tail or shrug of the skin. How is this essentially painless bite accomplished? I became familiar with the answer to this question when I did the research for my master's thesis, a microscopic study of the anatomy of the mouthparts of the female gallinipper, an exceptionally large North American mosquito.

Her mouthparts, virtually identical to those of most other female mosquitoes, are quite complex and marvelously adapted for making a painless entry into the skin. Being bitten by a mosquito is not at all like being pricked by a pin. The latter experience is painful, because the relatively dull pin rips our tissues as it forces them aside. But the razor-sharp mouthparts of the mosquito, drawn out into long, thin blades that fit together neatly to form the piercing-sucking apparatus, cut smoothly and carefully, like a hypodermic needle, as they gently insinuate their way into the skin. The maxillae slide on the other mouthparts, alternately shuttling forward and back as microscopic sharp teeth at their tips painlessly slice their way into the skin.

⁎ The largely sedentary larvae of black flies are aquatic—like the larvae of mosquitoes, but only more so. While most mosquito larvae must come to the surface to get oxygen directly from the atmosphere, the larvae of black flies have gills that can remove enough dissolved oxygen from the water to meet their needs. While mosquito larvae occur only in still water, black fly larvae occur only in flowing water. Some species can be found in the relatively gentle currents of large rivers, but most of them live in the turbulent and swiftly flowing water of mountain streams. Clell L. Metcalf, in a 1932 publication on the black flies of the Adirondack Mountains in New York, wrote:

> To find the young of blackflies one should go to the swiftest part of a stream, where the water churns or boils over stones, sticks, logs or other obstructions, or where vegetation such as the leaves of trees or of grasses or sedges breaks the surface of the current into ripples. Remove the stone or other object from the water and examine it in bright light . . . On the downstream side of the stones the dark gray to black appearing "worms" or "maggots" . . . will be found squirming or writhing slowly over the moist surface.

Living in swiftly flowing water offers two major advantages to black fly larvae. First, turbulent waters are well-aerated and supply the larvae with abundant oxygen. Second, the current brings food to these sedentary creatures. The black fly larva, attached to a rock or some other support by a sucker at its hind end and with its body hanging downstream with the flow, uses its fanlike mouth brushes to sift small bits of organic matter from the flowing water. The pupae of black flies live under the water in sac-like, silken cocoons that are usually attached to the same object to which the larva was attached. The adult emerges under water, rises to the surface, and then flies off.

Although black fly larvae are benign creatures, food for trout, the tiny adults, often only a tenth of an inch long, can be more troublesome than mosquitoes when they are abundant, as is often the case in spring and early summer in the mountains and in the northern forests. As Metcalf put it:

> The attack of black flies is totally different from that of mosquitoes. These flies hover about the body, being especially noticeable around the head. Their flight is swift and jerky but almost entirely noiseless. They can not be driven away by striking at them. They readily alight on skin or clothing and run greedily over the body like lice. Their bodies are firm enough so that they can squeeze under clothing and hair and feed on the tender protected skin, under conditions where mosquitoes would be crushed.
>
> The bite of the black fly is usually not painful at the time it is inflicted. One may watch a fly inflate its body with one's own blood without feel-

> ing any pain whatever; or see flies biting the forehead or eyelid of a companion who is unconscious of their presence . . . The pain, however, soon begins; and, depending upon individual susceptibility, the burning, aching and itching of the wound may continue for hours or even for days. Blood often flows from the wound, and individuals are sometimes seen with streaks of blood running down the face, arms or legs.

Some people have severe reactions to black fly bites. Swellings as large as marbles may form on the neck and face and may completely close one or both eyes. A few people may even run a fever, sometimes as high as 104°F.

Black flies get at the blood of their vertebrate hosts in a remarkable way, unlike that of any other insect. The blood is sucked up after the finely toothed mandibles snip through the skin like a pair of shears. The two mandibles overlap like the blades of a scissors and even have a pivot point that allows them to swing from side to side upon each other just as the blades of a scissors move on the screw that holds them together.

In late August of 1993, my friend Julia Berger and I were at Little Creek on the Delaware coast looking for the whiskered tern, a rare stray from across the Atlantic that was being reported on the rare bird alerts, the first bird of its species to be seen on this side of the Atlantic. We were a day late and saw neither it nor the white-winged tern, a more frequent stray from across the Atlantic that had been keeping company with it. But when we birded Bombay Hook National Wildlife Refuge later the same day, we were rewarded with good views of another stray from across the ocean, a sharp-tailed sandpiper, a lifer for both of us and the 600th species on my list of birds seen in America north of Mexico.

At least a dozen birders were there, their telescopes trained on the rare sandpiper and all of them slapping at the horse flies that were incessantly plaguing us. These flies, known as greeneyes or greenies along the Atlantic shore, are audacious biters, zooming in to land on some bare or thinly clad area of skin and inflicting a bite so painful that it instantly gets our attention.

Broadly speaking, we are likely to encounter two groups of horse flies in North America: the horse flies proper, which are in the genus *Tabanus,* and the deer flies, which are small horse flies of the genus *Chrysops*. Both belong to the family Tabanidae and are collectively referred to as tabanids or horse flies.

Except for the females in a handful of species, all female tabanids (in the broad sense) are blood feeders, but the males, which are seldom seen, drink only nectar and other plant exudates. The favorite targets of the females are large mammals, especially deer and cattle. Some species are known to attack reptiles such as lizards and sea turtles when the more preferred hosts are not available, but as far as I know tabanids have not been observed attacking birds. A few tabanids come out at night or twilight, but almost all of them are active only during the day. They generally prefer other mammals to humans, but, like the greenies at the shore, many if not most of them readily inflict their painful bites on people.

Horse flies, tabanids in the narrow sense, are generally larger than deer flies, and some of them can bite through thin clothing. Greenies are among the smaller horse flies, each being less than half an inch long, but most of the horse flies are much larger, and a common North American species, the black *Tabanus atratus,* is over an inch long and has a wingspan of about 2.5 inches. Although horse flies do occur in wooded areas, they are more likely to be found in open places than are deer flies. They usually attack the lower part of the body of any of their victims, especially the legs and ankles.

Deer flies are smaller than horse flies, most of them about the same size as a house fly. They occur mainly in open woodlands, the preferred habitat of the deer that are the preferred hosts of many of them. Deer flies, at least some of them, are all too willing to attack humans, and, size for size, their bites are as painful as are the bites of horse flies. They tend to bite on the head and neck. Often a dozen or more of them will buzz around your head, and if you are not wearing a hat they will land on your hair and try to wriggle down to the scalp. Insect collectors often capture these flies or try to get rid of them by swinging an insect net around their heads.

Many years ago I rode a horse into a woodland in Trumbull, Connecticut. Shortly after the horse and I had moved into the shade of the trees, I glanced down at the mare's ears, which were twitching frantically, and saw that they had turned from chestnut to yellow, because they were literally covered with yellow deer flies. Just about then, the horse, apparently tormented beyond endurance by the flies, bolted and ran out of the woods—

not along the trail but through the trees. I held on for dear life, bending over her neck to dodge branches that threatened to brush me out of the saddle.

The larvae of all tabanids are aquatic or live in wet soil. According to the late L. L. Pechuman and his coauthors, 109 species are known to exist in Illinois, but the larvae of only 77 of them had been described by 1983, and relatively little is known about the larval behavior and feeding habits of most of the known species. The larvae of horse flies are predators of other insects, but according to Pechuman, the feeding habits of the larvae of deer flies are poorly known. Some horse fly larvae secrete an oral venom that swiftly kills their prey, insects and even small vertebrates such as tadpoles or frogs. People who wade barefooted in fresh water are occasionally bitten by these larvae, as are Japanese farmers who wade in flooded rice fields, and as a result experience pain, severe itching, and swelling for as long as two days.

The periodical cicadas of Brood X, also known as the Great Eastern Brood, emerged in Vermilion County in central Illinois in the spring of 1987. At the height of the emergence, millions of these inch-long insects filled the air and crowded the trees, the singing of the males an all but deafening cacophony. A small flock of Mississippi kites, rare birds in central Illinois, appeared at Kennekuk County Park in Vermilion County to feast on the bounty. The news quickly spread on the birding grapevine, but I could not get to Kennekuk until the nineteenth of June. By then the emergence was dwindling, but a few cicadas still flew from tree to tree, and a few males still sang. Fortunately for me, the kites were still there, at least nine of them. I was elated; these were the first Mississippi kites I had seen in Illinois.

As I stalked and observed these birds, I heedlessly walked through knee-high grass in a shrubby area. A big mistake! Although I was wearing long pants, I had neither put on repellent nor tucked my pants in my socks. The next day my legs and waist were speckled with red spots. Chigger bites! Lots and lots of ferociously itching chigger bites.

The bites of these tiny mites are so itchy that it is almost impossible to keep from scratching them. The problem is nicely stated in a bit of doggerel quoted by the acarologist Tyler A. Wooley, who attributes it to a well-known entomologist, the late H. B. Hungerford of the University of Kansas:

The thing called a chigger
Is really no bigger
Than the smaller end of a pin.
But the bump that it raises
Just itches like blazes
And that's where the rub sets in.

Unlike the other bugs that eat humans, chiggers do not suck blood. They eat the cells of our skin. The females lay their eggs on the ground. The larvae that hatch from them crawl up onto vegetation and board a passing vertebrate if and when the opportunity presents itself. The almost microscopic larval chigger digs its sharp but minuscule mouthparts into the superficial layer of the skin and injects saliva that contains an enzyme that digests the skin cells, thus causing the formation of a long, narrow tunnel that extends down into the skin and is lined by hardened secretions that result from a defensive reaction of the host's skin. Contrary to what many people think, the chigger does not burrow into the skin. It remains at the upper end of the tunnel and ingests the partly digested cells that it contains. Also contrary to popular belief, chiggers do not die after feeding on a human.

Newly hatched chiggers, the only parasitic stage of the life cycle, attack not only humans but also a variety of other vertebrates. As a group, the various species of chiggers include most of the vertebrates among their hosts—rabbits, mice, squirrels, snakes, turtles, ground-nesting birds such as quail, and many others. After the young chigger has fed to repletion, it drops to the ground and undergoes four molts, passing through one feeding stage and two nonfeeding stages as it matures to the adult stage. The food habits of the adult and of the post-parasitic immature stages are not known, but Talmadge Neal and Herbert Barnett of the Walter Reed Army Institute of Research successfully raised an Asian chigger on a diet of springtail eggs. Springtails, as you will see in chapter 9, are tiny insects that live in the soil.

* Contrary to popular misinformation, the safest and surest way to remove an attached tick from your body is to grab the "head" as close to the skin as possible and pull gently but firmly so as to withdraw the mouth-

parts. Breaking them off in your skin is likely to lead to an infection. Tweezers are handiest but fingernails will do. Finding an engorged tick on your body—especially the more gluttonous female swollen to the size of a small grape with your blood—is unpleasant, to say the least. As Benjamin Yates, my eleven-year-old grandson, would say, "It's enough to gross you out."

All ticks, which are really very large mites, are blood feeders and attack only vertebrates. Some are quite host-specific, taking blood from only one or a few closely related species of mammals, reptiles, or birds; others, while confining their feeding to only one of these groups, range widely within it; and yet others include many species of birds, reptiles, and mammals among their hosts. As Burrus McDaniel pointed out, there is scarcely any domestic or wild animal that is not subject to their attack. And, as many birders have discovered, humans are often included.

Ticks can be much more than a nuisance. They transmit diseases of people and other animals and can in themselves cause a disease called tick paralysis, which is fairly common in wild and domestic animals but uncommon in humans. Tick paralysis of humans has occurred in various areas of North America but is most frequent in the northwestern United States and British Columbia—especially in small children, particularly girls with long hair. The paralysis progresses upward from the legs and can result in respiratory failure and death. A toxin released by an embedded tick, usually found on the scalp or the nape of the neck, is the cause, and the paralysis disappears within hours or days after removal of the offending tick.

Rocky Mountain spotted fever is a tick-borne disease that, despite its name, occurs throughout much of the United States and in some parts of Canada, Mexico, and South America. The bacterium that causes this disease is occasionally transmitted from wild rodents or rabbits to humans by ticks, the Rocky Mountain wood tick in western North America and the American dog tick in eastern North America. Fewer than a thousand cases are reported in the United States each year, but the mortality rate is high, between 5 and 10 percent. Tularemia, another bacterial disease of rodents and rabbits, can be transmitted to humans in several ways, but most commonly by the bites of ticks. The mortality rate is less than 1 percent and only about 200 cases are reported in the United States annually.

Lyme disease, by far the most prevalent arthropod-borne disease of humans in North America, is transmitted from small mammals and birds to people by a small acarine almost universally called the deer tick but known to acarologists as the black-legged tick *(Ixodes scapularis)*. (At the risk of offending my professional colleagues, I will bow to the popular consensus and refer to this creature as the deer tick.) Although Lyme disease has been known in Europe since the nineteenth century, it was first recognized in North America near Old Lyme, Connecticut, in 1975. But it was not until 1981 that the causative agent of this disease, a bacterium later named *Borrelia burgdorferi,* was discovered in a deer tick from Shelter Island, New York, by Willy Burgdorfer, a researcher at the Rocky Mountain Laboratory, National Institutes of Health, in Montana.

From 1982 to 1992, almost 50,000 cases of Lyme disease were reported in the United States, almost 10,000 cases in 1992 alone. Cases of local origin have been reported from all the states except Alaska, Hawaii, New Mexico, and Montana, but the incidence of Lyme disease is highest in the eastern states. Over 90 percent of the cases have occurred in the coastal states from Massachusetts south to Virginia. In 1991 the highest incidence was in Connecticut at a rate of 37.4 cases per 100,000 people.

According to the medical entomologist John F. Anderson, director of the Connecticut Agricultural Experiment Station in New Haven, the incidence of Lyme disease has been and continues to be on the increase. There are two major reasons for this. First and foremost, the population of white-tailed deer, the main dietary support of the deer tick population, has been on the increase—especially in recent years. Deer were almost exterminated from much of the United States during the nineteenth century. In 1896 the Connecticut Commission of Fisheries and Game reported that there were only a dozen or so deer in the state, but by 1988 the Connecticut Department of Environmental Protection estimated that the winter population was 31,000, and by 1996 it had grown to over 52,000. Second, humans are more exposed than ever to deer ticks because more people have moved into rural areas and because outdoor recreational activities are growing ever more popular.

The symptoms of Lyme disease occur in stages. The first stage, characterized by a spreading round or oval patch of red rash surrounding the site of the tick bite, occurs as the bacteria leave the area of the bite to invade the rest of the body. As the rash spreads outward, the center of the patch clears, leaving a ring of rash, often 4 or more inches in diameter, that surrounds

the original site of the bite. Later on, similar secondary patches of red rash may develop elsewhere on the body. The symptoms of the late stage of the disease are more serious and may include chronic arthritis and cardiac and neurological problems.

Although the blood of deer is the dietary mainstay of the deer tick population, these ticks have been found feeding on 120 different animals, including 52 mammals, 60 birds, and 8 reptiles, all lizards. Adult females feed mainly on white-tailed deer, but will also attack other large and medium-sized mammals such as striped skunks, foxes, wolves, coyotes, and black bears. Immatures occasionally feed on deer or other large mammals, but most often attack small mammals, ranging from shrews to mice, and birds, including wrens, vireos, warblers, and sparrows. Only immatures have been found on lizards. So far, the bacterium that causes Lyme disease has been isolated from 11 mammals and 9 birds in the northern United States. White-footed mice are particularly important reservoirs for this disease, but white-tailed deer do not support the bacteria in their bodies and are thus not a part of the disease reservoir, although they are virtually the only source of blood for adult deer ticks.

The deer tick occurs throughout the eastern half of the United States, from Maine and Florida west to Texas, Kansas, and Minnesota. It also occurs in the eastern provinces of Canada and is common at Long Point, a favorite spot for Ontario birders. (Other species of ticks transmit Lyme disease in the west.) In the northern part of its range, this tick requires two years to complete a generation. Tiny "seed ticks," six-legged and loosely known as larvae, hatch from eggs in August and early September; seek hosts; and when fully engorged, fall to the ground, where they survive the winter in the humus or perhaps in the moist soil in an animal burrow. The following spring they molt to become eight-legged nymphs. The nymphs board a host, engorge in 3 to 5 days, and then fall to the ground and molt to the adult stage. Mating may occur on vegetation, but most often takes place on the host, where the adult female feeds from 9 to 11 days. The fully engorged female drops to the ground and ultimately lays from 1,000 to 3,000 eggs in the humus layer of the soil.

It is not easy for a wingless and slow-moving deer tick, or most other kinds of ticks for that matter, to find and board a host animal. It is largely a matter of chance. The hungry tick climbs to the tip of a blade of grass or onto the leaf of an herb or shrub and waits for a suitable animal to come along. When it perceives the odor of a potential host, it goes on the alert, waving its front legs and reaching out as far as it can. If the animal comes close enough, the tick grabs hold of fur, plumage, or clothing, crawls to a suitable place on the animal's body, and embeds its mouthparts deep into the skin. Recurved hooks on the embedded mouthparts prevent the tick from losing its hold and falling off before it has satiated itself with blood.

Entomologists take advantage of this host-finding behavior of ticks to capture them and thus estimate their population size. This is done by walking through tick habitat as you drag close behind you a rectangle of flannel cloth. The ticks, alerted by the odor and carbon dioxide released by your body, grab onto the fuzzy cloth and hold on tightly. John Bouseman, an authority on ticks, tells me that this collecting technique captures only ticks that are willing to take blood meals from humans. For example, an abundant Illinois tick that attacks raccoons but not humans is never taken in this way. Furthermore, this technique captures the adults but never the immature forms of certain ticks that feed from humans only after they have molted to the adult stage. It seems that the ticks recognize the odor of their proper host.

It may take a deer tick several days to engorge itself with blood. Males take relatively small meals, but a female may suck up as much as 300 times her own weight in blood. Since its bite is painless, a tick often goes undiscovered for the 10 days or so that it takes to finish feeding, more than enough time for the transfer of the Lyme disease pathogen, which occurs after 24 to 48 hours.

Although scorpions, spiders, and some insects such as bees and wasps do not consume parts of us, they can nevertheless be a nuisance or even a serious threat to us, because they can sting or otherwise inject venom. Arthropod venoms vary in their effects on humans according to the sensitivity of the person and the potency of the venom. A few people develop an allergy to certain arthropod venoms and suffer severe reactions—even life-threatening anaphylactic shock—as the result of a sting. But most people are not allergic, and their response to a venom is commensurate with its toxicity. The sting of a scorpion may be as deadly as the bite of a venomous

snake. But the venom of the little, brown sweat bees that drink perspiration from our skin and occasionally sting us when we disturb them causes only mild to barely perceptible pain.

Scorpions, predaceous relatives of the spiders, use the stingers at the end of their abdomens primarily to kill their prey but also to defend themselves when they feel threatened. In North America they occur in the southern and western states, with one species extending as far north as the southern tip of Illinois, but the most dangerous species are found in the Southwest. The toxicity of scorpion venom to humans varies greatly. The stings of some species cause only discomfort, but the stings of others can be lethal. During a 20-year period early in this century, scorpions of the genus *Centruroides* killed twice as many people in Arizona as did all other venomous animals combined. The sting of the Durango scorpion of Mexico, another member of the genus *Centruroides,* is often fatal, especially to children. During a 37-year period, 1,608 children in the city of Durango died from its sting.

It is best to avoid all contact with scorpions. When you are in scorpion country, do not poke your bare fingers under rocks or into piles of firewood. When camping, shake your shoes out before putting them on in the morning.

Some spiders—but not all—subdue their prey by injecting them with venom. But the great majority of spiders—even the big, dangerous-looking ones—never or almost never bite people. One that sometimes does is the brown recluse spider, which is found in the central United States and Arizona and California. It is most likely to be encountered indoors, where it sometimes bites if disturbed, although it may also bite when disturbed outdoors by a person turning over a rock or tilling a garden. But the dangerous spider that we are most likely to encounter outdoors in North America is the black widow, which occurs in southern Canada and throughout the lower 48 states. Only the web-spinning female, black and with a red hourglass marking on her underside, is likely to bite. The bite may cause no more than two or three days of discomfort, but has been fatal in about 4 percent of the reported cases.

As Maurice James and Robert Harwood related in *Medical Entomology,* about half of the recorded black widow bites are on the genitalia of men or boys who use outdoor privies. Female black widows spin webs across the hole in the seat—a good place to catch passing flies—and bite foreign objects that dangle in the web, perhaps mistaking the foreign object for prey or, more likely, seeing it as a threat to their egg sac. People are also bitten when they poke their fingers into places where the spiders lurk—perhaps when picking berries or tomatoes or when gathering firewood. When I picked up the skull of a cow in a pasture in Texas, I found that its interior was inhabited by several black widows with egg sacs and newly hatched spiderlings. I was not bitten, but might have been if I had poked my fingers into the eye sockets rather than picking up the skull by the horns.

The fear of wasps and hornets—sometimes a well-founded fear—can be traced back to biblical times. When the Israelites were making their way from Egypt to the promised land, Jehovah made a covenant with them, promising that "I will send my terror before thee, and will discomfit all the people to whom thou shalt come, and I will make all thine enemies turn their backs unto thee. And I will send the hornet before thee, which shall drive out the Hivite, the Canaanite, and the Hitite from before thee" (Exodus 23:27 and 28). According to Roger Akre and his coauthors, there are about 15,000 species of stinging wasps in the world, about 4,000 of which can be found in North America. The great majority of them are solitary (nonsocial) species that, with some exceptions, use their sting only to kill or paralyze the insects or other arthropods that they feed to their larvae. (The adults generally drink nectar and juice from their prey.) The stingers of wasps, like those of their ant and bee relatives, are modified ovipositors,and consequently only the females can sting.

Solitary wasps will probably sting if you grab one of them in your fingers, but otherwise they are rarely aggressive, seldom sting, and with some notable exceptions, their venoms are not particularly painful to people. The giant cicada killer—an adult female may be 2 inches long—is a good example of an innocuous solitary wasp. In July and August the females cruise trees to capture dog-day cicadas that they paralyze with a sting and place in

burrows in the soil as food for their larvae. Although they are huge and dangerous looking with their yellow- and black-barred abdomens, I have never been stung by one of these inoffensive creatures, not even when I hunched over their burrows with my camera and even went so far as to snatch cicadas away from females so that wasp and cicada would not disappear down the burrow before I was ready to photograph them. I was not stung even when I took the same cicada away from the same female several times in succession.

But the social wasps, known as yellowjackets or hornets, are different. They seldom sting when they are away from the colony, but they quickly respond to a disturbance of their nest, flying out en masse to sting the intruder. A colony consists of a queen plus from dozens to hundreds or even thousands or workers, all females and all capable of stinging repeatedly. (In temperate climates, the stingless males do not appear until late summer or fall.) Not only do social wasps sting en masse and repeatedly, but their venom, unlike the venom of most solitary wasps, causes severe pain in humans, and presumably other mammals, because it contains compounds, such as histamines and serotonin, that cause pain when they are injected subcutaneously. A stinging wasp braces itself with all six legs as it thrusts the stinger into the skin. The European hornet, introduced into the New York area between 1840 and 1860 and now found in much of southern Canada and the eastern half of the United States, braces itself similarly but gains even firmer purchase by clamping onto the skin with its mandibles. It seems likely that social wasps evolved pain-causing venoms and their bellicose defensiveness as a response to raids on their colonies by mammals such as mice and skunks. Nevertheless, some birds and even animals as large as bears find wasp colonies, which may contain hundreds or even thousands of juicy grubs and pupae, to be a tempting food resource.

People are sometimes stung because they accidentally bump into the aerial nest of a colony of black and white bald-faced hornets. Their large, football-shaped nests, familiar to most people, are usually suspended from a branch of a tree or shrub and are made of paper that the wasps produce by mixing their saliva with vegetable matter, often wood fibers that they gather from rotten wood or tear from a weathered log or fence post. People are even more likely to unintentionally step into the hidden nest of a yellowjacket, the name usually applied to the smaller social wasps of North

America. Yellowjacket nests, which are usually subterranean, consist of a large paper structure not unlike the nest of a bald-faced hornet.

According to Bert Hölldobler and E. O. Wilson, there are about 8,800 known species of ants worldwide, plus an immense number that remain to be discovered. All known ants are social. Many of them sting in defense of their colonies—not surprising since the ants are descended from wasps—but many other species of ants have lost their stings. Nevertheless, most stingless ants are eminently capable of defending their colonies. Some have a soldier caste with huge, powerful, and wickedly sharp mandibles. Others can cause a sharp stinging pain, despite the absence of a stinger, by abrading the skin with their mandibles and then squirting formic acid into the wound.

About 580 species of ants occur in America north of Mexico, but only a few more than two dozen of them are likely to sting people: several species of harvester ants and the fire ants that you met in a previous chapter. All the ants in both of these groups have viciously painful stings. When a fire ant is about to sting, it sinks its powerful mandibles into your skin for leverage and then arches its abdomen as it sinks its stinger into your flesh. A sharp pain is felt almost instantly, the area immediately surrounding the site of the sting soon reddens and swells, and the site of the sting ultimately becomes a pus-filled lesion that may persist for days. Both fire and harvester ants attack en masse. A person may receive hundreds of stings that will have a seriously debilitating effect. Fire ants have killed domestic animals and, on rare occasions, even people. Harvester ants are equally threatening. David Costello mentioned two human deaths from harvester ant stings that were reported in Oklahoma, and described what happened after he intentionally permitted one of these ants to bite and sting the back of his hand: "The immediate pain was like the stab of a wasp. Within ten minutes an inflamed circle of skin appeared. A white spot remained in the center of the one-inch red area for more than two hours. The pain and itching subsided in one hour, but beads of perspiration appeared in the area of the sting for more than a week. . . . It is easy to believe that the southwestern Indians once staked their human victims on ant hills for torture."

Fire ants are found mainly in the southeastern United States, including

A birder in a field dotted with fire ant nests

Texas. Only one species of harvester ant occurs in the Southeast, but about 20 other species occur in the western half of the country, as well as in British Columbia.

The way to avoid being stung by either fire ants or harvester ants is to stay away from their colonies. This is not hard to do, because the above-ground features associated with their underground nests are large, highly visible, and easily recognized.

Fire ants, which occur in pastures, crop fields, and other open areas, build an earthen mound that may be 30 or more inches in diameter and from 15 inches to 3 feet tall. The mound is as hard as a rock, but inside it is honeycombed with tunnels. Fire ants are omnivorous and eat vegetation as well as insects and other animals. Harvester ants build flatter and wider mounds that may have a basal diameter of from 2 to 5 feet but are seldom more than a foot tall. They surface their mound with particles of the material available, sand in desert areas, bits of lava in volcanic areas, and sometimes more unusual materials. According to David Costello, "harvester ants near Grand

Junction, Colorado, use petrified shark teeth that have remained in conglomerate [rocks] since the Cretaceous period, and in the Red Desert of Wyoming they bring up small rubies from below ground . . . [which] can be seen shining red in the sunlight." These ants clear and maintain an extensive bare area, from 10 to 50 feet in diameter, surrounding the mound. True to their name, they harvest seeds that they store in the nest, but the bare area is not a result of the harvest. It is created by purposeful clearing. The harvester ants cut the plants surrounding the nest into small pieces and discard them at the edge of the clearing or let the wind blow them away.

All bees collect nectar and pollen as food for themselves and their young. Social bees, mainly the honey bees and the bumble bees, accumulate large stores of these nutritious substances. These stores plus the many larvae and pupae in the nest are a bounteous food resource for mammals ranging from mice and skunks to bears and humans. Consequently, social bees, like social wasps, have evolved effective defensive behaviors and venoms that cause severe pain. The solitary bees, which by far outnumber the social species, accumulate fewer resources and are consequently seldom the target of raids by mammals. They are less defensive than the social species and their venoms generally cause far less pain.

Like the social wasps, the social bees seldom sting when they are away from their nest. But if their nest is disturbed, the angry workers swarm forth to sting the intruder. Their venoms, like those of the social wasps, include substances that cause severe pain.

Bumble bees nest in a burrow in the soil, often in the abandoned home of a mouse. People are most often stung as a result of stepping on the nest. Captive honey bees nest in hives, but feral colonies nest in above-ground cavities of all sorts, such as hollow trees and the space between the inner and outer walls of a building. The European varieties of honey bees, the ones that are kept in the United States, are not particularly aggressive. They almost never sting when away from the nest and even at the nest seldom attack unless the nest is severely disturbed.

But Africanized honey bees are a different story. Honey bees of the African race are fiercely defensive. They attack en masse with little or no provocation. Their introduction into the New World, well chronicled by Mark L.

Winston, has proved disastrous. In 1956, with the approval of the Brazilian authorities, Brazilian scientists imported a large number of African queens. Their intention was to hybridize the African race, good honey producers, with their own domestic bees, which were of the less productive but also much less aggressive European race, to produce a more productive variety of honey bee. In 1957, 26 African queens, accompanied by swarms of European workers, escaped from an experimental apiary near Rio Claro, a town about a hundred miles north of São Paulo. The escapees thrived, interbreeding with honey bees from feral colonies and even with domestic bees from commercial hives. Soon virtually all the feral honey bees in Brazil were "Africanized," and they started to spread to the south and the north, expanding their range 100 to 200 miles every year.

Africanized bees, sometimes called killer bees in horror films and the tabloid press, have now overrun all but the cold southern tip of South America and all of Central America and Mexico. There have been numerous reports of unprovoked attacks on animals and humans in these areas. Hundreds of people, thousands of domestic animals, and countless wild animals have been killed by them. Africanized bees reached the Rio Grande Valley in southern Texas in the fall of 1990. Not long after, they became established in southern Texas, New Mexico, Arizona, and California. There have been several attacks on people, and one of these attacks, on an old man already weakened by disease, was fatal. The infested area is subject to a limited quarantine to slow the bees' northward progress. No one knows how far north they can spread, but since they are native to the warm temperate areas of southern Africa and have established themselves in warm temperate areas of South America, it is possible that they will eventually occupy all or most of the southern United States.

According to Sue Hubbell, bee keepers have learned to control domestic colonies of these vicious bees by not placing hives near houses, by wearing protective clothing, and by being generous with applications of the smoke that soothes the bees. But there is, nevertheless, a serious threat from wild colonies that may be encountered by birders or other people who frequent the outdoors.

* A few caterpillars inject venom even though they lack a stinger. They bear sharply pointed, venom-filled hairs. If you brush against one of these

caterpillars, the hairs—known as urticating hairs—penetrate your skin and release their venom when they break off. Urticating caterpillars occur throughout most of the United States, but the nastiness of their venom varies with the species. Saddle-back caterpillars cause severe pain. In *The Moth Book,* W. J. Holland wrote of this caterpillar: "The green caterpillars with their little brown saddles on the back are familiar to every Southern boy who has wandered in the corn-fields, and many a lad can recall the first time he came in contact with the stinging bristles as he happened to brush against the beastie. Nettles are not to be compared in stinging power to the armament of this beautifully colored larva."

The venom of caterpillars of flannel moths is far more painful and even dangerous. In South America I have seen flannel moth caterpillars that are at least 2 inches long and half that wide. They are so densely covered with long, white or yellow hairs that neither head nor legs are visible. One of them sitting on the upper side of a leaf looks like a toupee for a tiny elf. Fortunately, I never brushed against one of these caterpillars, but I heard about an American biologist who did. The pain was immediate and so excruciating that he had to be hospitalized. But he suffered no permanent harm and was released after a few days.

# People Fight Back

# 8

European ladies of the fifteenth to eighteenth centuries wore flea traps between their breasts. The trap, which hung from a ribbon around the neck, was a hollow cylinder—sometimes carved of ivory—perforated by holes big enough for a flea to pass through. Inside was a rod coated with blood or some sticky substance that was supposed to entangle any fleas that entered. These traps may have worked, but I doubt it. I cannot imagine why a hungry flea would have entered one. After all, the trap was surrounded on all sides by soft skin that had only to be pierced to obtain the sought-for blood meal. Both James Busvine and Brendan Lehane have described these traps and some others that might possibly have worked. These other traps, known as *Flohpelze* (flea furs) in German, consisted of the pelt, or some part of the pelt, of a small animal that hung from a chain and was often decorated with a jeweled artificial head such as the one illustrated in E. Steingraber's *Antique Jewelry. Flohpelze* may, as Lehane suggests, have been the forerunners of the fur stoles that were still fashionable early in this century. These furry traps may have worked. A hairy pelt might actually have distracted and temporarily detained fleas searching for a blood meal.

Cleanliness has all but eliminated the human flea *(Pulex irritans)* from our homes, mainly by eliminating the debris in which the larvae live. The fleas that sometimes infest modern homes and only occasionally bite us are almost certain to be cat fleas *(Ctenocephalides felis),* which, notwithstanding their name, attack both dogs and cats and prefer both of them to humans. Larval cat fleas live in the pet's bedding or in the pile of the carpet the pet sleeps on. Thus a modern birder is not likely to harbor fleas when she goes out with her binoculars, and surely not the human flea that tormented women of previous centuries. But when medieval birders went afield—the word birder then referred to men who trapped or shot birds—they probably brought along human fleas from home in their clothing.

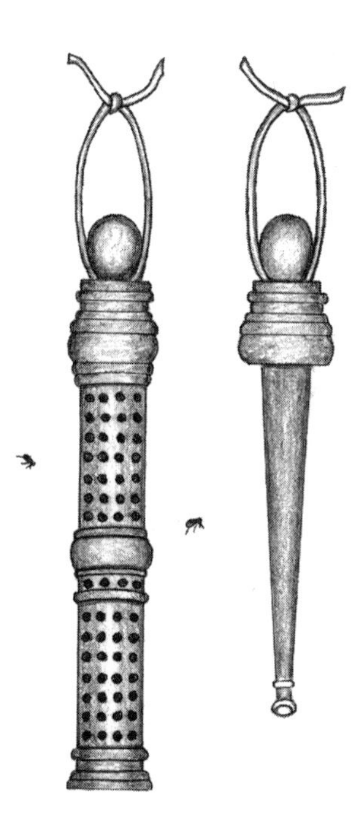

A seventeenth-century flea trap with its sticky inner rod removed and shown on the right

In those days, fleas were everywhere, even present as uninvited guests at the fashionable salons held by prominent ladies of the day and attended by artists, writers, and other notables. In late sixteenth-century France, Madeleine des Roches and her stepdaughter Catherine held such a salon at Poitiers. During a gathering in 1579, a flea appeared on the bosom of the young and nubile Catherine. The gentlemen who were present were enchanted and wrote verses to commemorate the event. Etienne Pasquier, a minor French poet, wrote with envy of the fortunate flea that stung the tender flesh between the breasts of Catherine, "the most beautiful of the beautiful." These poems belong to a genre of flea-on-bosom poetry that began in antiquity, continued well into the nineteenth century, and even today is echoed in jokes and limericks. Many of these poems express envy of the flea, as did Pasquier's, and some speculate, more or less naughtily, about which part of the woman's body the flea might visit next.

According to *Le Grand Larousse,* the comprehensive French encyclopedia, both Madeleine and Catherine died of the plague at Poitiers in 1587. We know, therefore, that the famous incident at the salon was not Catherine's last encounter with a flea. The bacterium that causes bubonic plague is

transmitted from rats to humans by the bites of fleas—not by human fleas but by other species that regularly bite rats and occasionally bite humans.

The third volume of Edward Topsell's 1658 *History of Four-Footed Beasts and Serpents and Insects,* actually written by Thomas Muffet (also spelled Moffet, Mouffet, and Moufet), is the often-quoted *Theater of Insects,* which first appeared in Latin as *Theatrum Insectorum* in 1634. Muffet's writing style and spelling are decidedly old-fashioned, but are no more difficult to understand than are the works of Shakespeare. But when we consider what he had to say and his approach to the attainment of knowledge, we find ourselves on unfamiliar and long since abandoned intellectual ground. He believed, for instance, in the spontaneous generation of life, as did Aristotle. Muffet wrote that the origin of fleas "is from dust, chiefly that which is moystned with mans or Goats urine." Noting that people who wash and change their clothes seldom have lice, while those who neither wash nor change their clothes usually do have lice, he concluded that lice, which "all . . . breed from humours," arise from "sweat corrupted," the stale sweat on unwashed clothing. "Hence it is that Armies and Prisons are so full of Lice, the sweat being corrupted by wearing alwaies the same cloathes, and from thence ariseth matter for their original by the mediation of heat."

Implicit in Muffet's discourse on lice is the good advice that cleanliness, especially frequent changes of clothing, is the key to freeing the body of these parasites. But the real reason for the soundness of this advice is far different from his conjecture that lice arise from "corrupted sweat." While head lice stick their nits (eggs) to the hair, body lice lay theirs in the clothing. If clothing is frequently washed, the nits are destroyed before they can hatch. Thus reproduction becomes impossible and the louse infestation soon disappears. In their *Introduction to Entomology,* first published in 1815, William Kirby and William Spence, English entomologists, recount another and far less pleasant way of eliminating lice from the body that "Hungarian shepherds find effectual to put to flight these insects [fleas] and their neighbours the lice. This is not, as you may be tempted to think, by a remarkable attention to cleanliness.—Quite the reverse.—They grease their linen with hog's lard, and thus render themselves disgusting even to fleas."

In Muffet's day, European science was in its earliest infancy. He was, after all, a contemporary of the philosophers who initiated scientific inquiry, seeking knowledge by looking at the world rather than by following the scholastic approach of endlessly poring over accepted dogmas, the Scriptures, and the writings of Aristotle and a few other ancients. Among Muffet's contemporaries were Galileo, forced by the Church to recant his belief that the earth is not the center of the universe; William Harvey, who first demonstrated that blood circulates in the body; and Francesco Redi, designer and performer of an experiment that disproved the then current notion that maggots appear spontaneously in putrefying meat. Muffet was, in a small way, an observer of nature, but he was essentially a product of his time and by no means a practitioner of the scientific method.

The conflict between dogma and rationalism that persists to this day is succinctly expressed by a question, revealing although humorously put, that Frank Cowan quotes in his 1865 *Curious Facts in the History of Insects:*

> In the Gentleman's Magazine for 1746, there is a curious letter on a "certain *creature,* of rare and extraordinary qualities"—a Louse, containing many humorous observations on this "*lover* of the human race" and concluding with some queries as to its origin and pedigree. "Was it," the writer asks, "created within the six days assigned by *Moses* for the formation of all things? If so, where was its habitation? We can hardly suppose that it was quartered on *Adam* or his lady, the neatest, nicest pair (if we believe *John Milton*) that ever joyned hands. And yet, as it disdained to graze the fields, or lick the dust for sustenance, where else could it have had its subsistence?"

Much remains to be learned about fleas and all the other creatures of this world, but we have come a long way since Muffet's time. I think that he would be astonished and probably enthralled by what we now know about fleas, especially by the marvelous story, largely uncovered by Miriam Rothschild, of how rabbit fleas synchronize their reproductive cycle with the reproductive cycle of the female rabbit. The female rabbit's cycle is triggered and guided from ovulation to birth by sex hormones that she produces in her body. The female flea's reproductive cycle is controlled by the same hormones, the very rabbit hormones that she imbibes with the rabbit's blood. According to Rothschild, the flea's ovaries do not mature so long as she takes her blood meals from a male rabbit or a female that is not pregnant. But when she feeds from a pregnant female, whose blood is loaded with hormones, among them estrogen, her ovaries begin to develop. When

the young rabbits are born, the fleas transfer to them, and only then does the female flea mate with the male. She then lays her eggs in the nest with the young rabbits, at the only time and in the only place where her larval offspring are guaranteed a salubrious home with a constant supply of their food, organic particles shed by the rabbits and flecks of dry blood that result from excess rabbit blood that adult fleas squirt out through the anus as they feed.

✱ On the subject of mosquitoes (sometimes known as gnats in England), Muffet wrote of the experiences of some of the early explorers of North America, stories that are reminiscent of more recent tales of the mosquitoes of the New Jersey or Texas shores.

> The English *Gnats* are not so stinging as others, nor do they raise so great pimples, but the lesser sort of them is the more cruel, and yet they leave nothing behind them but a little itching spot, like a flea-biting. The *Gnats* in *America* . . . do so slash and cut, that they will pierce through very thick clothing . . . people when they are bitten will frig and frisk, and slap with their hands their thighs, buttocks, shoulders, arms, sides, even as a carter doth his horses. The gnats about *Terra incognita* or *New-found-land,* and *Port Nicholas,* as also in divers other Northern parts, are to be seen in great numbers, and of extraordinary bignesse, as the sea-men and *Olaus Magnus* affirm. The cause of their multitude *Cardanus* attributes to the unintermitted heat and the length of the day. The cause of their bignesse to that watery and unctuous moisture which was gotten together by reason of the long cold.

In considering "the generation of gnats"—the birth of mosquitoes—Muffet does not invoke spontaneous generation. Nevertheless, Muffet's ideas about how mosquitoes originate seem ridiculous in view of our present knowledge about how they procreate: "Of the Generation of Gnats Natures secretaries do diversly dispute: *Albertus* saith their material is watery vapors. *Aristotle* denies the Gnats should be generated from Gnats unless by means of a little worm as Flies are." Aristotle was right as far as he went. But Muffet argued that "since that they [the gnats] do not use copulation, I do not perceive how that can be." He then went on to present another view, "that

Gnats do come of certain worms breeding in wood, when as yet every man knowes that Gnats are produced of worms in the *Navew, Privet, Mastick, Turpentine, Wilde Fig-tree,* and other like trees."

Among the methods of warding off the attacks of mosquitoes that Muffet describes are two that make an interesting contrast, a workable one devised by the pragmatic Greeks, no doubt based on observed reality, and a useless one that he attributes to the English and that was conceived in ignorance and obviously not evaluated by observation. "The Grecians," he wrote, "have devised a kinde of tent or covering in manner of a net, of linnen, woollen, or silk; which being hung about their dining rooms and beds, kept the Gnats from entering in." He then goes on to describe another method, surely ineffective but presumably current in the England of his day. "Our Countrymen that live about the Fens have invented a canopy . . . with less cost but the same profit, which they call a *Fen-canopy,* being made of a broad, plain, half dry, somewhat hard piece, or many pieces together of Cowes dung, and these they hang at their beds feet: with the smell and juice whereof the Gnats being very much taken and feeding thereon all the night long, let them sleep quietly in their beds without any disturbance or molestation." The cow dung must have attracted hordes of filth flies, as dung- and carrion-feeding flies are known, perhaps including some small ones that vaguely resemble mosquitoes, but it was surely ignored by blood-seeking female mosquitoes. The users of fen canopies never thought to get at the heart of the matter by counting mosquito bites so that they could compare how often people "protected" by cow dung were bitten with how often people not so protected were bitten.

Netting is still widely used: screens prevent flies, mosquitoes, and other insects from invading our homes; hats draped with netting protect our heads from the attacks of mosquitoes and black flies; tents and campers are provided with screening; and mosquito nets are still used to cover beds. But since World War II, chemical repellents have been the most widely used form of personal protection against biting insects, mites, and ticks.

The development of effective insect and mite repellents that can be applied to the skin was a high priority during World War II. In the Pacific theater of operations, according to William Horsfall's *Medical Entomology,* more soldiers were killed or disabled by diseases transmitted by arthropods than by enemy action. Horsfall relates that at one site in eastern New

Guinea, Allied troops suffered multiple infections of malaria, 4,000 cases per 1,000 men per year of this mosquito-borne disease. In other words, each person was, on average, infected and reinfected with malaria four times. Scrub typhus, transmitted by a species of chigger mite, was second only to malaria in disabling troops in the Pacific. Effective synthetic mosquito and mite repellents were developed in time to help alleviate these problems, and even more effective repellents, which work against ticks, mites, and blood-sucking insects such as mosquitoes, black flies, and horse flies, have been developed since.

The earliest repellents, known for centuries, were natural plant products. They were not very effective against mosquitoes, but they did give limited protection from some other biting insects. Oil of citronella, extracted from a south Asian grass, is the active ingredient in some "natural" repellents that are now on the market and in the citronella candles that are burned outdoors to chase away insects. There was also oil of pennyroyal, extracted from aromatic European and North American mints. Its repellent properties have been known for a long time. In his 1758 *Systema naturae,* which marked the beginning of the modern science of taxonomy, Carolus Linnaeus named the North American species of pennyroyal *Hedeoma pulegioides,* which translates from the Latin as sweet-smelling shield against fleas.

Most of the repellents now on the market contain Deet, the common brand name for N,N-dimethyl-meta-toluamide, as their active ingredient. Among them are the various formulations of Cutter and Off! Deet applied to the skin gives good protection against mosquitoes, black flies, punkies, horse flies, other biting flies, chiggers, and ticks. How long the protection lasts varies with the insect in question: more than six hours against mosquitoes but only about three and a half hours against stable flies, according to *Consumer Reports* (December 15, 1993).

But using Deet has its risks. It is absorbed through the skin, readily enters the bloodstream, and on rare occasions causes adverse reactions—especially in children. Since it takes as long as two months to eliminate Deet from the system, repeated applications over a long period can result in the accumulation of a possibly dangerous concentration in the body. Among the adverse reactions to Deet are skin rashes, muscle cramps, nausea, irritability, lethargy, brain damage, and even death.

What can you do to minimize this risk? You can switch to repellents that contain no Deet, such as those that contain citronella or other natural substances. Some people swear by these products, but others have found them to be far less effective than Deet. It is a good idea to use Deet in moderation. Check the labels and avoid repellents that contain more than 30 percent Deet—no more than 10 percent for children; apply it no more often than necessary; wash it off when you come indoors and no longer need it; and, as the labels warn, do not apply it to the hands of children. Little hands go too often into the mouth, and Deet should not be taken internally. Perhaps the best way to minimize your risk is to wear long, lightweight pants and long- sleeved shirts and to spray the repellent on them rather than on your skin. That way little or no Deet will get into your bloodstream. Furthermore, Deet remains active far longer when applied to cloth than when applied to skin.

You can avoid chiggers by staying out of the weedy, grassy areas where they lie in wait to board your body. Stay on the path and do not brush against nearby vegetation. If you plan to enter weedy areas, you can keep chiggers off by tucking your pants into your socks and spraying your ankles, beltline, and fly with a repellent. If you do get chigger bites, they will heal sooner if you do not scratch them. I find that the itching is relieved—and the urge to scratch curtailed—by the application of Lanacane or some other nonprescription cream that contains a contact anesthetic such as benzocaine.

In 1955, the World Health Organization (WHO) began a worldwide campaign to eradicate malaria by preventing contact between humans and the *Anopheles* mosquitoes that transmit the several forms of this disease. (There are four forms of human malaria caused by four different species of *Plasmodium*). People are most likely to be bitten by anophelines as they sleep during the night in homes that are not protected by screens, which is often the case in developing countries. As part of the WHO project, the mosquitoes, which rest on walls and ceilings during the day, were killed by spraying these surfaces with DDT, which left a toxic residue that persisted for as long as six months and that was absorbed through the mosquitoes' feet.

For a time, the eradication campaign was a great success. Malaria was eliminated from 36 countries, and its rate of occurrence was reduced in many other countries. A 1991 publication of the National Academy of Sciences edited by Stanley Oaks, Jr., and others reported that malaria had been all but eliminated from Sri Lanka by the 1970s. But today over a million cases are reported from this island nation each year. There have been massive resurgences of malaria in India, Pakistan, and elsewhere. In 1995, WHO reported that worldwide there are now 300 to 500 million cases per year and from 1.5 to 2.7 million deaths per year, the great majority of them in Africa. Malaria, the most important of the arthropod-borne diseases, is again at least as prevalent as it was before 1955. The failed eradication campaign ended in 1969, but WHO still supports some efforts to reduce mosquito populations.

Why did the eradication campaign fail? The basic reason, as Robert Metcalf has explained, is that *Anopheles* mosquitoes became resistant to DDT and the other inexpensive insecticides that were first used to control them. (Mosquitoes and other insects continue to become resistant to newly developed insecticides.) Different insecticides were substituted, but the mosquitoes soon became resistant to them and to many others as one failed insecticide after another had to be replaced. The campaign became more and more costly because each substituted insecticide was usually more expensive than the one it replaced. Substituting Malathion for DDT increased costs by about fivefold, and then switching to propoxur increased costs by about twentyfold. As Metcalf pointed out, "This combination of resistance and economics is responsible for the return of epidemic malaria." To make matters even worse, the malaria-causing plasmodia have, in some areas, become resistant to chloroquine and other antimalarial drugs.

Malaria is steadily increasing among tourists and other travelers who visit malarial areas of the tropics. Anyone who plans to visit such areas should consult a physician about taking preventive doses of antimalarial drugs and should also prepare to protect himself or herself against mosquitoes. Contracting malaria in the United States is possible although unlikely. Before World War II it was common in the southern states, but since 1950 there have been only two dozen or so outbreaks in this country, all minor, the majority occurring in California. In the summer of 1988, there was an outbreak of malaria in San Diego County, California, 30 cases, a small outbreak but the largest to occur in the United States since 1952. Since then, there have been a few more cases in the San Diego area. These infections probably originated with infected migrant workers who came to California from places where malaria is permanently established. Judy Hansen, superintendent of the Cape May County Mosquito Extermination Commission in New Jersey, tells me that there were two cases of malaria in New Jersey in 1991, and two cases in Florida during the first half of 1996.

Using repellents, spraying your backyard with an insecticide before a barbecue, and other individual efforts are fine as far as they go, but the best and most economical way to control mosquitoes is to stop them at the source, to prevent larvae from maturing to become biting adults. Military people understand this concept. Just as it is more efficient to bomb a weapon-producing factory than to cope with the weapons one by one after they have been deployed over a large area, it is more efficient to control mosquito larvae in the relatively limited aquatic environments in which they breed than to contend with the adults after they have dispersed over an area that will probably be thousands of times more extensive than are the combined breeding sites of the larvae. Since mosquitoes do not respect property lines, their control is, generally speaking, most economically and efficiently accomplished if there is a community-wide or even more broadly based effort in which everyone "chips in"—usually via a local system of taxation—to support the experts and workers who design and implement the control program.

Cape May, New Jersey, is justly one of the most famous birding localities in the United States. In the spring, flight-weary songbirds of many species crowd the few remaining trees and shrubs. In the fall, migrant songbirds are again abundant and a stately procession of hawks, especially sharp shins, soars by. Like the rest of the New Jersey shore, Cape May is also famous for the ferocious mosquitoes that breed in the vast salt marshes behind the sandy barrier islands on which lie the sea beaches. Mecca for the sun worshipers and other holiday visitors who invade each summer, these beaches may also be invaded by salt marsh mosquitoes, and according to Thomas Headlee, these annoying pests often move 35 to 40 miles inland and have appeared in areas as much as 100 miles distant from the marshes in which

they originated, "infesting seriously more than one-half the state's land surface and annoying very seriously nearly three-fourths of her population."

In her 1990 presidential address to the annual meeting of the American Mosquito Control Association, Judy Hansen quoted eighteenth-century accounts of the tormenting salt marsh mosquitoes of New Jersey. "The flat and marshy parts of the state . . . which are very numerous are infested with myriads of mosquitoes which give intolerable annoyance to man and beast," wrote a Philadelphia physician in 1792. A few years earlier another observer had written of marshes "covered with stagnant waters which infest the air, and give birth to the mosquitoes with which you are cruelly tormented; and to an epidemical fever which makes great ravages in summer; a fever known likewise in Virginia and the southern states." The fever was surely malaria, not transmitted by *Aedes sollicitans,* the most abundant of the salt marsh mosquitoes in New Jersey, but probably transmitted by the most important vector of malaria in the eastern United States, *Anopheles quadrimaculatus,* a less numerous marsh breeder. But in the eighteenth century, the relationship between mosquitoes and malaria, first announced by Ronald Ross in 1897, was not yet known. Another eighteenth-century physician, quoted by Thomas Headlee, wrote of an experience on the marshy New Jersey coast: "Surrounded by millions of musquetoes we were obliged to spend the time until daybreak on the deck of the little vessel . . . I had covered myself with a cloak, and a thick sail, and the night being extremely warm I suffered as in a perfect sweat-bath, but the musquetoes found their way through."

Today the salt marsh mosquitoes of New Jersey are under control. They are still there, but seldom do we see the unrestrained outbreaks that were common until the mid-1960s. All but one of the 21 counties in the state are now served by mosquito extermination commissions or agencies, including all 11 coastal counties. "Extermination commission" is a misnomer that dates back to the optimistic and ecologically less sophisticated days of the early twentieth century. Salt marsh mosquitoes could be exterminated only by destroying the marshes, and no one wants to do that. Their loss would be an environmental catastrophe.

Organized efforts to control New Jersey's salt marsh mosquitoes began in the first decade of this century, when the state legislature granted John B. Smith, an outstanding entomologist of his day, a small sum of money to investigate the problem. His studies centered on the mosquitoes, because he realized that a rational control plan could be devised only in the light of a thorough understanding of their biology, life history, and habits. In Smith's day, the field of insect control had not yet been compromised by the advent of DDT and the other synthetic organic insecticides with their false promise of a quick fix without the need for understanding the pest insects and their ecology. Insecticides will always have a role in insect control—an important role—but they are not a substitute for an informed and rational approach to the problem.

*Aedes sollicitans,* the most numerous and important of the salt marsh mosquitoes in New Jersey, is actually a floodwater species that, like *Aedes vexans,* lays its eggs not in water but on the soil in places that are subject to periodic flooding. After taking a blood meal, the female lays her eggs on the moist mud of a depression that is high enough not to be flooded by daily tides but that will be flooded by storms or the spring tides that occur at the times of the new and full moon. If the depression remains dry, the eggs will survive without hatching for months or even years. When the depression is flooded, the eggs hatch within minutes or hours. After completing the larval and pupal stages, often in less than a week, the adult females move inland in search of a blood meal so that they can, in turn, return to the marsh to lay their eggs.

Smith's rational plan for the control of salt marsh mosquitoes, based on his understanding of the habits of the mosquitoes and the ecology of the salt marsh, was reiterated by Thomas Headlee in 1945:

> To produce mosquitoes, warm water free from killifish must lie on the marsh surface long enough for maturity to be reached or mosquitoes will not be produced.
>
> Briefly stated, the control of the salt marsh mosquito is a matter of so ditching the marsh that none of the water which remains upon it will stagnate, but all will rise and fall with the tide and be everywhere penetrated by killifish. The only exception to this rule is the permanent salt marsh pool which, being constantly stocked with killifish, [produces] no mosquitoes.

Killifish *(Fundulus heteroclitus),* also called top minnows and known to me as mummies when I was a child on the Connecticut coast, are ravenous devourers of mosquito larvae.

The ditching of the marshes, planned and implemented by engineers

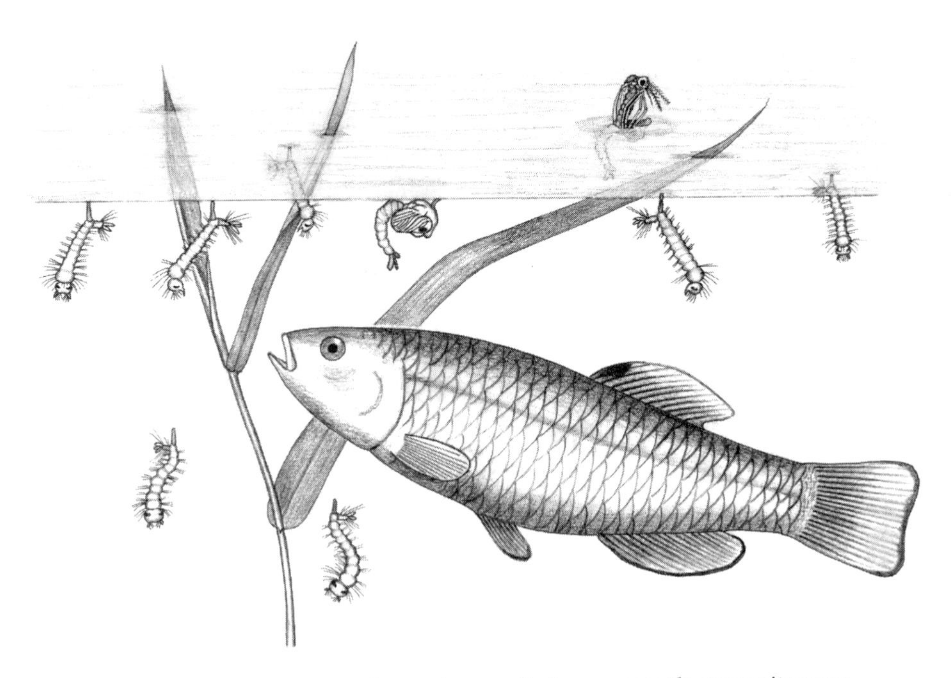

A killifish about to snap up one of several mosquito larvae; note the mosquito pupa just above the fish's head and near the upper right an adult mosquito emerging from the pupal skin

rather than entomologists, was accomplished mainly during the Great Depression of the 1930s as a make-work project of the Civilian Conservation Corps (CCC) and the Works Projects Administration (WPA). It was a failure. The engineers, more concerned with the ditching itself rather than with its intended purpose, laid the ditches out in a standard rectilinear grid of evenly spaced ditches. A ditch was dug where the engineering plan dictated even if it did not intercept a tidal pool. Pools were left unserved if the plan did not call for a ditch to drain them. Although mosquito populations were somewhat reduced, the mosquito problem continued to be severe, because too many breeding sites in the marsh were neither drained nor connected with the creeks or deep pools that are the permanent homes of killifish.

Just this August, a friend and I visited Judy Hansen, director of the Cape May County Mosquito Extermination Commission, at her headquarters in the town of Cape May Court House. After Judy filled us in on the commission's control program, George Conover took us up in a helicopter and gave us a bird's-eye view of the salt marshes from an altitude of about 200 feet.

No rectilinear grid of ditches was to be seen. Those that we saw would not satisfy anyone's craving for geometrical orderliness. They had been laid out according to the principles of open marsh water management, an ecologically based technique for controlling salt marsh mosquitoes that was originally developed through the cooperation of the Cape May and Cumberland County mosquito control commissions; the New Jersey Division of Fish, Game, and Wildlife; and entomologists at Rutgers University. The radially arranged ditches connect shallow, mosquito-breeding depressions with more permanent bodies of water such as tidal creeks, deep salt marsh ponds, or reservoirs 3 feet deep that were scooped out if no creek or pond was nearby. When the shallow depressions are dry, mosquitoes lay eggs in them. The eggs hatch when the depressions are flooded, but the larvae never mature because they are eaten by killifish that swim into the depressions via the radial ditches that connect with the creeks, deep ponds, or reservoirs that harbor permanent populations of these fish.

Open marsh water management has resulted in a significant reduction in the amount of insecticide that is applied to the salt marsh. Breeding sites that have been ditched produce no mosquitoes, but the relatively few sites that have not yet been ditched can produce them and must sometimes be treated with an insecticide that kills mosquito larvae—a readily biodegradable insecticide that does not accumulate in the environment and that is not concentrated in food chains. These potential breeding sites are under close surveillance and are not treated unless significant numbers of mosquito larvae are actually present. The sites are reached by helicopter, and are checked for larvae by the tried-and-true method of scooping up samples of water with an ordinary dipper mounted on a long handle.

As we flew over the marsh we saw great blue herons, egrets, and little blue herons dotting the marsh wherever there was water. A northern harrier glided just above the grass as it searched for rodents. Shorebirds walked muddy shores along tidal creeks, and flocks of gulls and terns stood on sand bars at the mouths of creeks. Later in the year, George told us, the marsh would be host to great flocks of snow geese, brants, and ducks. It is a healthy environment that supports an abundance of life ranging from crickets, mud snails, and diamondback terrapins to rails, ospreys, meadow voles, and foxes. As a nursery for blue crabs, shrimps, and many different kinds of fish, such as flounder, black drum, weakfish, bluefish, and striped bass, the marsh remains a vital link in the marine ecosystem.

# A Brief Guide to the Insects

# 9

Insects are surpassingly numerous and wonderfully diverse. There are at least 900,000 known species worldwide. It is not easy to know the 805 species of birds that occur in America north of Mexico, although it can be done. But no one can possibly learn to know the over 88,000 species of insects that occur in the same area. How could anyone ever comprehend and understand so many? The answer is, of course, to classify them, to arrange them systematically in groups of similar and related species. Biologists classify the million plus species of known animals in a hierarchy of groups, each group more inclusive than the preceding group and each consisting only of species that descended from a common ancestor. I will begin this chapter at the top with the most inclusive group, the phylum, and then move down the hierarchy to the less inclusive groups.

The insects constitute one subdivision, the class Hexapoda of the phylum Arthropoda. Students sometimes tell me that technical terms, such as the two that I just used, are all Greek to them. They are right—at least concerning these two terms. Both come from Greek roots, and both are informative once we look them up in a dictionary to find out what their roots are. *Hexa* means six, and *poda* means leg, hence six-legged, a notable distinguishing characteristic of insects. Together *arthro,* a joint, and *poda* mean jointed legs, one of the major features of the animals known as the arthropods.

The phylum Arthropoda and 35 other phyla, such as those that include the sponges, mollusks, and vertebrates, make up the animal kingdom. The arthropods, which, in addition to the insects, include the centipedes, millipedes, arachnids (spiders and their relatives), and crustaceans (lobsters and their relatives), have in common the basic anatomical plan of an external skeleton, a segmented body, and paired, segmented legs.

The major subdivisions, called classes, of the phylum Arthropoda are evolutionary variations on the theme of this basic plan. Thus the classes are distinguished by the various ways in which the segments of the body have become grouped to form distinctive body regions and according to the different evolutionary fates of the legs, of which there was presumably one pair per segment on the bodies of the earliest arthropods.

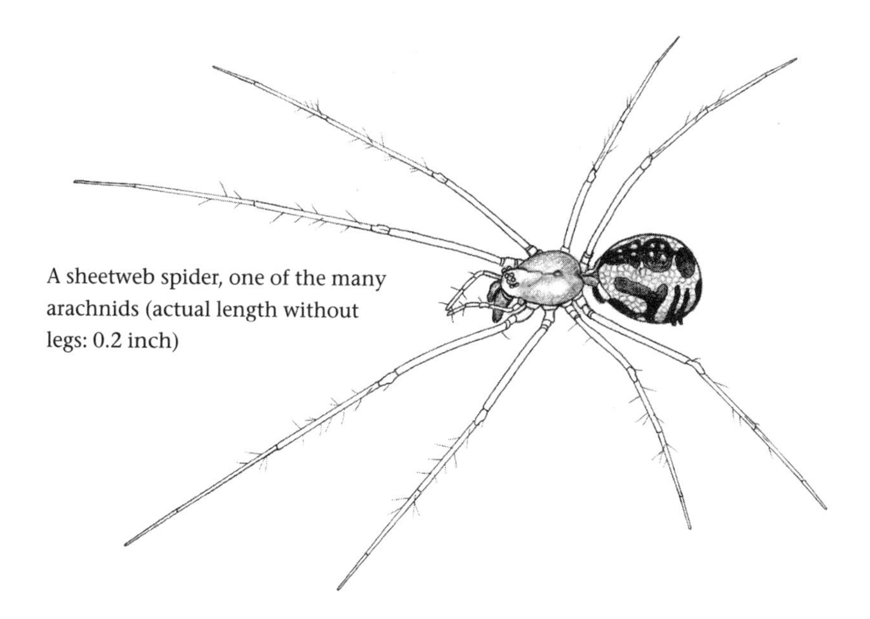

A sheetweb spider, one of the many arachnids (actual length without legs: 0.2 inch)

All animals in the class Arachnida (Greek *arachna,* a spider), which includes spiders, ticks, mites, scorpions, and others, are arthropods that have the segments of the body grouped to form two regions, an abdomen and a cephalothorax (Greek *cephala,* head). They lack antennae, and all the legs of the abdomen have been lost. Two of the original 12 legs of the cephalothorax are modified to form a pair of antenna-like pedipalps, and two others form a pair of mouthparts that may be specialized as fangs. The remaining 8 appendages of the cephalothorax retain their original function as legs. Most arachnids prey on insects and other small animals, but some suck blood from vertebrates, and many of the mites feed on plants.

Crabs, lobsters, crayfish, shrimps, fairy shrimps, sowbugs, and their relatives constitute the class Crustacea (Greek *crusta,* a crust, referring to the hard shell that covers most crustaceans). They are mostly aquatic and may be predators, herbivores, or scavengers. Almost all have only two body

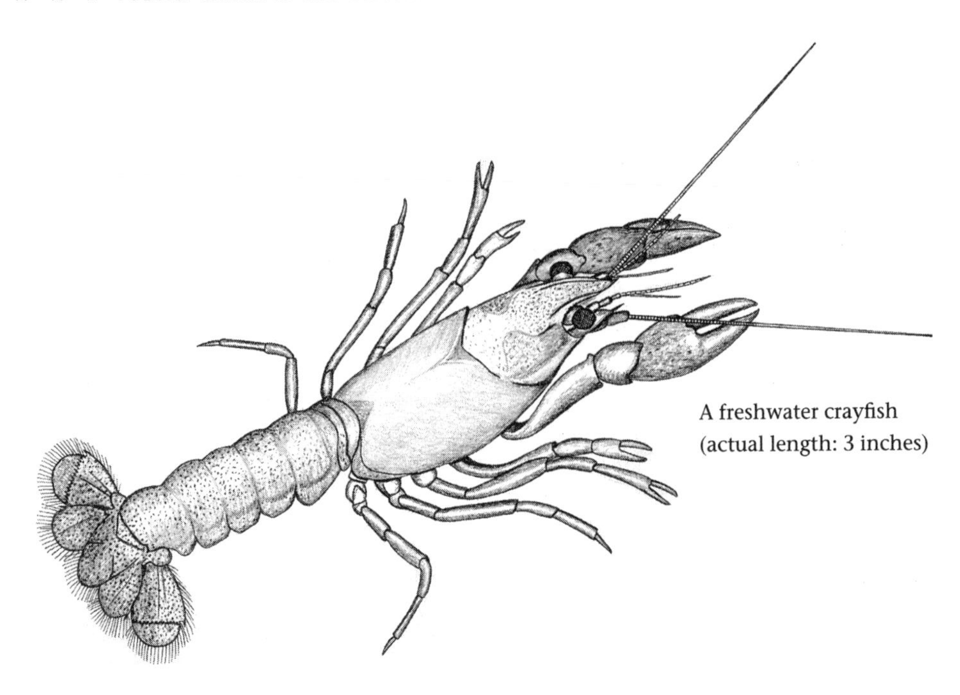

A freshwater crayfish (actual length: 3 inches)

regions, a cephalothorax and an abdomen. The cephalothorax bears two pairs of antennae, legs modified as mouthparts, and five pairs of walking legs. Some or most of the legs on the abdomen have been retained and are often modified as swimmerets.

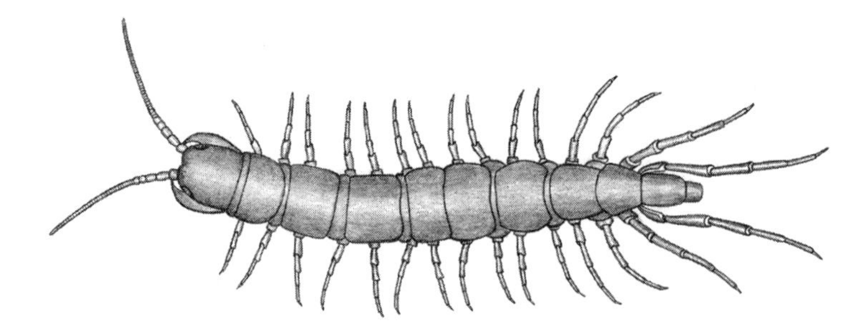

A centipede (actual length: 1 inch); note the poison "jaws" slung forward under the head (at the left)

In the predaceous, terrestrial centipedes, class Chilopoda (Greek *chili,* lip), the segments of the body have become grouped as two body regions: the head, with a pair of antennae and legs modified as chewing mouthparts, and the trunk, a series of similar segments, each of which bears a pair of walking legs except for the first segment. The legs of this segment have been modified as venom-injecting "jaws" that are slung forward under the head and look something like a pair of lips, hence the scientific name of this class.

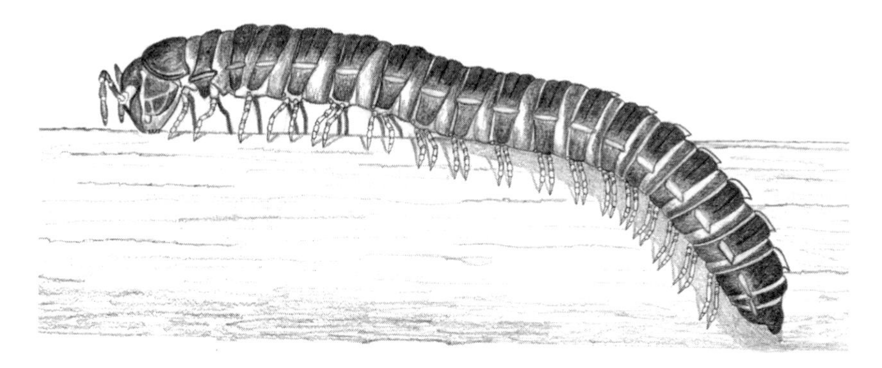

A millipede (actual length: 1 inch)

The largely plant-eating and always terrestrial millipedes, class Diplopoda (Greek *diplo,* doubled), are similar to the centipedes except for the absence of the venom jaws and the seeming presence of two pairs of legs per segment when a millipede is viewed from the side or from above. If we flip a millipede onto its back, we can see that each seemingly two-legged segment actually consists of a pair of segments, each bearing one pair of legs and with each pair of segments covered by one dorsal plate.

Insects, the members of the class Hexapoda, are arthropods that have the segments of the body grouped to form three body regions: head, thorax, and abdomen. On the head they have a single pair of antennae and mouthparts that are actually modified legs. They have a pair of legs on each of the three thoracic segments but none on the abdomen, and they usually have wings on the thorax.

✱ The importance of flight to the insects is reflected by the central role that the wings play in their classification. The insects are divided into two groups, subclasses, according to the presence or absence of wings. The subclass Apterygota (Greek *a,* without, and *pteron,* wing) includes just over 7,000 primitively wingless species, less than 1 percent of the known insects.

They all lack wings and are descended from ancestors that never evolved wings. The other subclass, the Pterygota (Greek, with wings), includes all the winged insects and some secondarily wingless ones, such as fleas and lice, that are descended from winged ancestors but lost their wings as an accommodation to a specialized life style, such as living as parasites among the feathers of a bird or in the fur of a mammal. The subclass Pterygota includes over 99 percent of the known insects. At least as judged by numbers of species, the winged insects have been evolutionarily vastly more successful than the primitively wingless insects.

The subclass Pterygota is, in turn, divided into two groups according to the structure of the wings. The group called Paleoptera (Greek *paleo,* ancient), the most primitive of the winged insects, includes only the mayflies (Ephemeroptera) and the dragonflies and damselflies (Odonata), the only orders of winged insects that have not evolved a mechanism for flexing the wings down flat against the body. When not flying, Paleoptera can only hold their "old-fashioned" wings together vertically up over the back or straight out to the side like an airplane. Consequently, they are limited in their activities, precluded from burrowing or entering tight, narrow places.

The Neoptera (Greek *neo,* new), all of the remaining insects, evolved the hinges and muscles required to fold the wings down flat over the back and thus get them out of the way so that the insect can retain the ability to fly and yet enter cracks and crevices, burrow in the soil, or tunnel in plant tissues . The small size of all insects and these capacities of the Neoptera in particular are major reasons for the extraordinary evolutionary success of the insects. The Neoptera (over 880,000 known species) have been far more successful than the Paleoptera (less than 7,000 known species).

The Neoptera can be subdivided into two groups according to what type of metamorphosis they undergo. One group consists of the orders that have *gradual metamorphosis,* and the other includes the orders that have *complete metamorphosis.* Judging by the number of species that have evolved in these two groups, complete metamorphosis is the more useful and adaptable strategy for survival. The insects with complete metamorphosis outnumber those with gradual metamorphosis by more than five to one.

The insects are classified in 31 orders. I discuss 19 of them here, those orders whose members are associated with birds or are likely to be encountered by birders. The others are small orders whose members are seldom seen. The 19 orders discussed here include 98 percent of all the known insects. Frank B. Gill, in his recently published text on ornithology, divides the birds among 29 orders. It is a curious fact that entomological taxonomists accommodate almost a million species in 31 orders, and that it takes almost as many orders to classify only 9,617 species of birds. Insects are certainly more diverse than birds. Thus it seems that an order of birds may not be equivalent to an order of insects in the breadth of the diversity that it includes. Different authorities may recognize different numbers of orders of insects, birds, or other animals, defining them more or less broadly so that they include more or fewer species. For example, the orders Phasmida (walkingsticks), Orthoptera (grasshoppers, crickets, and katydids), Mantodea (mantises), and Blattaria (cockroaches) were once lumped as a broadly defined order termed Orthoptera, and the current order Phthiraptera (lice) is a combination of the once more narrowly defined orders Mallophaga (biting lice) and Anoplura (sucking lice).

The brief summaries that follow will give you a nodding acquaintance with the major orders of insects, enough so that you can recognize at least some of their members and so that you know at least a bit about how some of them live.

## Subclass Apterygota (Primitively Wingless Insects)

### Order Collembola (Springtails)

Springtails, constituting the order Collembola, are very abundant, but they are seldom seen, because they are all tiny and because most of them live hidden in the soil, leaf mold, rotting logs, or similar locations, where they eat decaying vegetation, algae, fungi, or other organic substances. One North American species, *Podura aquatica,* lives on the surface of ponds and lakes, sometimes occuring as highly visible gray rafts of thousands, hundreds of thousands, or even millions of individuals. Springtails derive their common name from a forklike appendage at the tip of the abdomen, which, when cocked under the abdomen and released with force, propels the insect high into the air. Many authorities believe that the Collembola are not insects and place them in a class of their own.

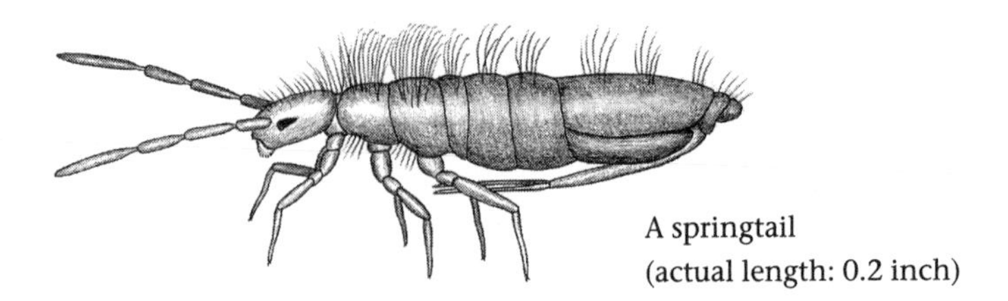

A springtail
(actual length: 0.2 inch)

**Order Thysanura (Bristletails)**

The order Thysanura consists solely of the bristletails. These insects are named for the three long bristly tails at the tip of the abdomen. Most bristletails are found in caves, in debris, under rocks, or in ant nests, but the two best-known species, the silverfish and the firebrat, are common residents of human dwellings, usually benign presences, although they may nibble on the bindings of books to get the starchy glue. Both of these household insects are carrot-shaped, whitish or grayish-brown, covered with minute scales, and rather slinky and fast moving.

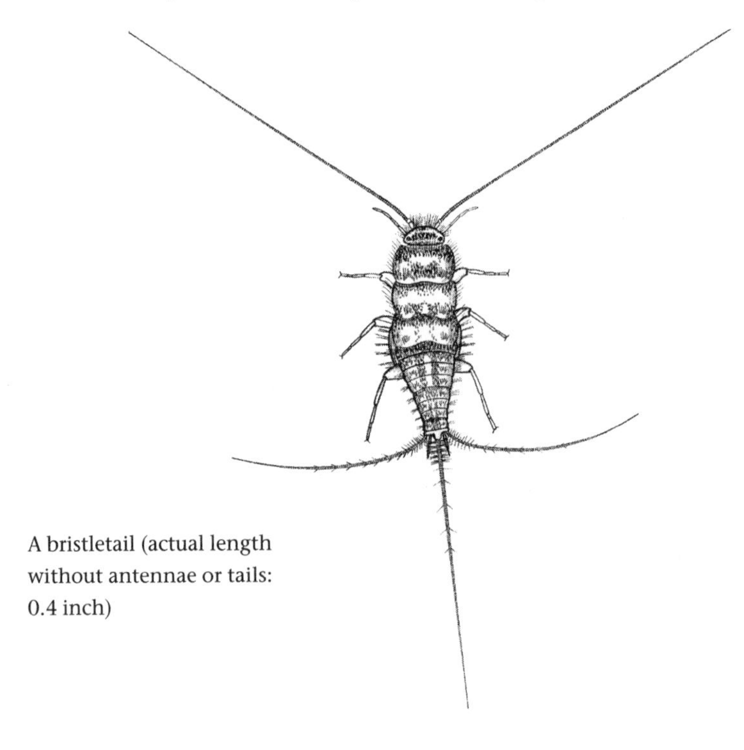

A bristletail (actual length without antennae or tails: 0.4 inch)

## Subclass Pterygota (Winged and Secondarily Wingless Insects)

### The Paleoptera (Primitive Winged Insects)

**Order Ephemeroptera (Mayflies)**

The aquatic nymphs of the mayflies, the only insects in the order Ephemeroptera, generally eat algae or organic debris and have three long tails at the tip of the abdomen and a row of leaflike gills along each side of the

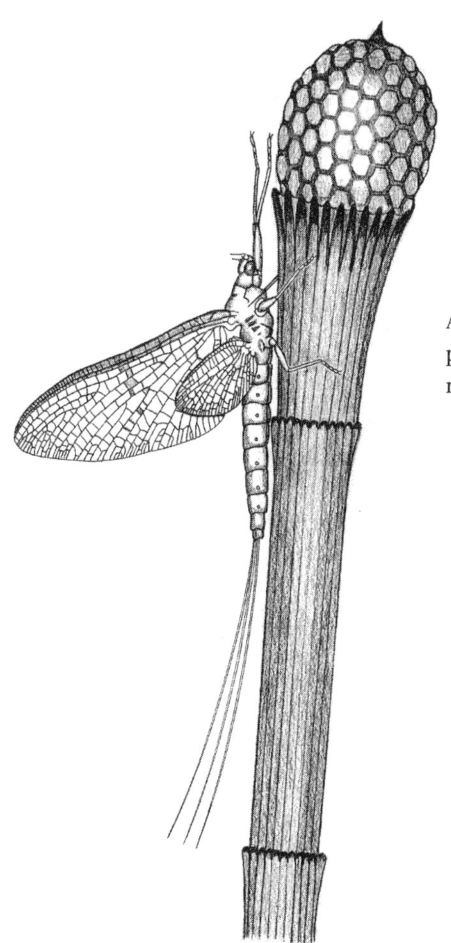

A mayfly perched on a horsetail, a primitive plant (actual length of mayfly without tails: 0.6 inch)

abdomen. True to their scientific name, the winged adults are ephemeral, not capable of feeding and seldom surviving for more than a day. The males have two "penises" to match the two "vaginas" of the females, an anatomical characteristic almost unique among insects. Adults of a given species emerge from the water synchronously, thus facilitating the finding of a mate. Immense, night-flying swarms occur along the Great Lakes and the rivers of the Mississippi drainage system. In 1954, I saw such an emergence along the Rock River in Janesville, Wisconsin. Countless millions of mayflies, attracted to street lights and brightly lit store windows, beat themselves to death and accumulated on the street in great piles that were as much as 4 feet deep. The next morning trucks and a front-end loader were required to clean up the mess.

**Order Odonata (Dragonflies and Damselflies)**

The dragonflies and damselflies, the two kinds of insects in the order Odonata, are predaceous both as nymphs and adults. The aquatic nymphs use their elongated, prehensile lower lips (labia) to grab insects and even small fish. The adults, accomplished aerialists despite their "old-fashioned" wings, catch flying insects in a "basket" formed by their hairy legs. Damselfly nymphs have three leaflike external gills at the tip of the abdomen. In dragonflies the gills are internal; they line the muscular rectum and are bathed by water that is sucked in and squirted out through the anus. When a dragonfly nymph is in a hurry, it can forcefully expel water through its anus and thus speed through the water like a rocket. Dragonflies and damselflies have a unique way of copulating. The male's genital opening and his genital claspers are at the tip of his abdomen, as is the case with other insects, but he inseminates the female by means of a secondary set of genitalia at the base of his abdomen. He first ejaculates into his secondary genitalia and then uses the secondary "penis" to pass his semen into the "vagina" at the tip of the female's abdomen, thus freeing the genital claspers at the tip of his abdomen for a different and very important function, clasping his mate behind the neck so as to monopolize her. In many species, he will hold onto her for the rest of the day as she flies about laying her eggs, thus preventing her from mating with other males and assuring that the eggs that she lays that day will be fertilized by his sperm. Some male damselflies use their secondary intromittent organ to remove and discard the semen of a preceding male from the female's sperm pouch before mating with her.

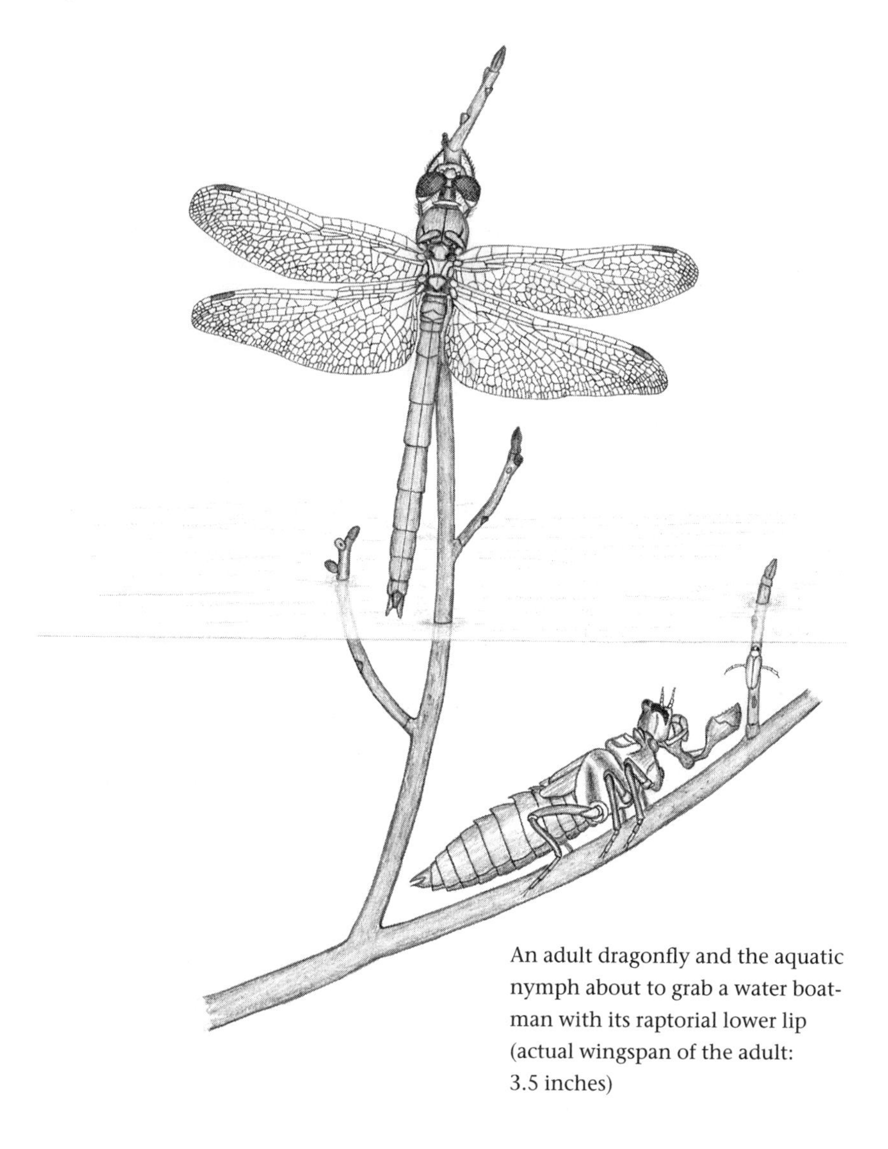

An adult dragonfly and the aquatic nymph about to grab a water boatman with its raptorial lower lip (actual wingspan of the adult: 3.5 inches)

## The Neoptera (Higher Winged Insects)

### ORDERS WITH GRADUAL METAMORPHOSIS

#### Order Phasmida (Walkingsticks)

The large, sluggish insects in the order Phasmida, collectively known as walkingsticks, are leaf feeders that live mainly on trees. Except for one species in Florida, all North American walkingsticks are wingless. Some tropical species are disguised as leaves, but all the North American ones have thin, elongated bodies that make them look like twigs, a camouflage that has been experimentally shown to deceive insect-eating birds.

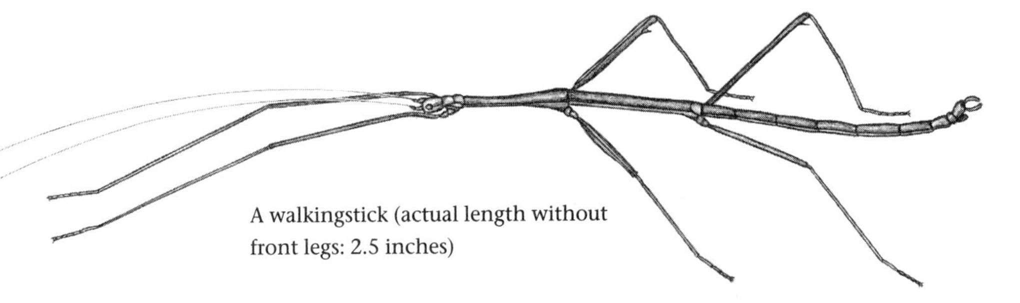

A walkingstick (actual length without front legs: 2.5 inches)

#### Order Orthoptera (Grasshoppers, Crickets, Katydids, and Others)

The members of the order Orthoptera, which includes the grasshoppers, crickets, and katydids, are of medium to large size and can be recognized by their jumping hind legs, their leathery front wings, and the membranous

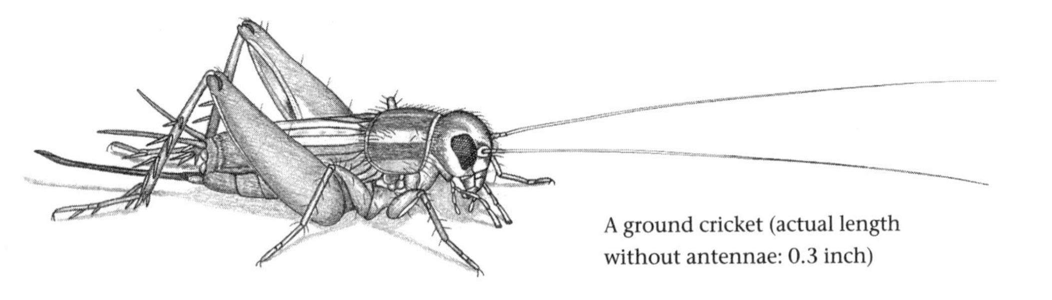

A ground cricket (actual length without antennae: 0.3 inch)

hind wings, which are folded like a fan beneath the front wings. Grasshoppers feed on plants; crickets are mainly omnivores; and katydids are mostly plant feeders, although some are omnivorous or predaceous. The Orthoptera are notable for the sounds that they make to attract mates and for the great swarms of grasshoppers (locusts) that migrate through North Africa and the Near East, eating everything green in their path.

**Order Mantodea (Praying Mantises)**

Praying mantises, the sole insects in the order Mantodea, are predators, mainly of other insects, and are easily recognized by the elongated first segment of the thorax and by the pair of raptorial legs that it bears. The

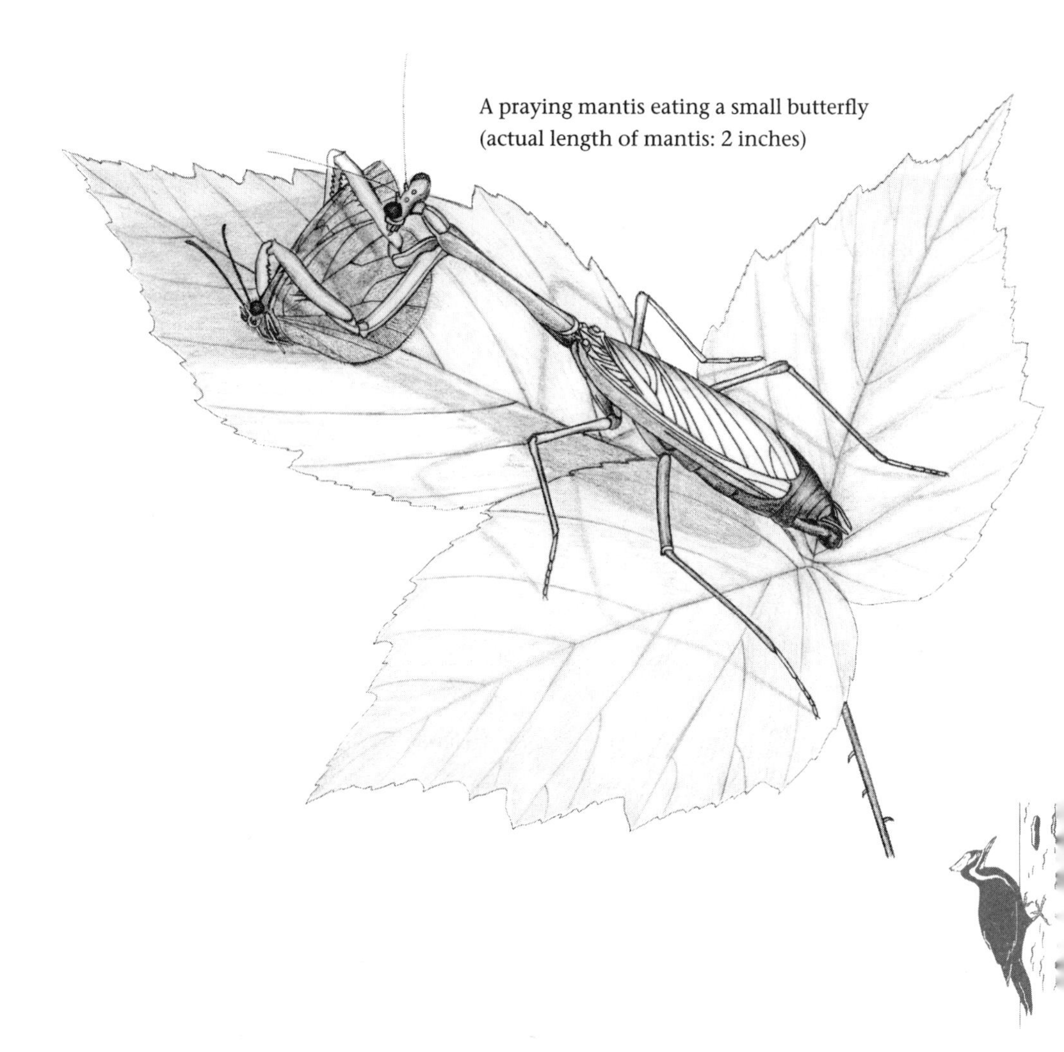

A praying mantis eating a small butterfly (actual length of mantis: 2 inches)

large, light tan egg cases of the introduced Chinese mantis, often attached to dead stems of goldenrods or other tall herbaceous plants, are now a feature of the winter landscape of eastern North America. Our native species produce somewhat smaller but less conspicuous egg cases. Mantises are ambushers, and are usually camouflaged so as to deceive their prey. Some tropical species are brightly colored but blend in with the colorful flowers on which they wait motionless for prey insects, such as bees, to appear.

**Order Blattaria (Cockroaches)**

All members of the order Blattaria are cockroaches. They are distinguished by their flattened bodies, long antennae, and the expanded dorsal shield of the first thoracic segment, which usually covers the head. They are mainly tropical, but there are several native North American species, known as wood roaches, that live in leaf litter and rotting logs. The most familiar of our cockroaches are the few species that live as obnoxious and unwelcome

A cockroach (actual length without antennae: 0.5 inch)

guests in our homes. These pests are really tropical and in our climate survive only in heated buildings. Little is known about the feeding habits of most cockroaches, but the pest species are omnivores.

**Order Isoptera (Termites)**

The termites, sometimes known as white ants, constitute the order Isoptera. All termites are social and, much like ants, live in large colonies whose members are divided among specialized castes: small, white, wingless sterile workers; wingless sterile soldiers with greatly enlarged heads and mandibles; fully winged primary reproductives; and often supplementary reproductives with only partially developed wings. The front and hind wings of termites are, unlike the wings of all other insects, similar in shape, texture, and venation. All castes include both sexes, unlike the castes of social ants,

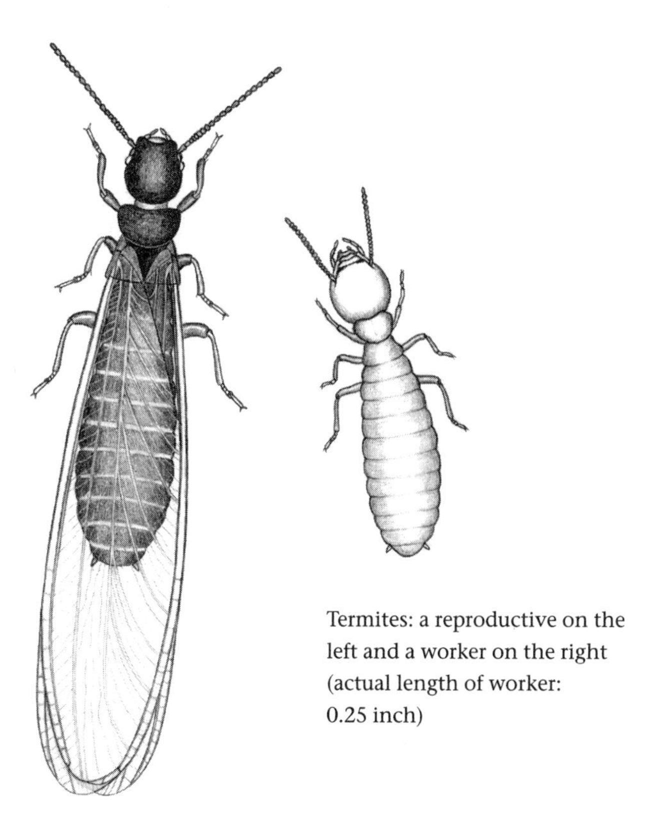

Termites: a reproductive on the left and a worker on the right (actual length of worker: 0.25 inch)

wasps and bees, in which workers and soldiers are all females. New colonies are founded by pairs of flying primary reproductives, which seek out a suitable site and then snap off their wings as an accommodation to life underground. North American termites eat dead wood. They cannot make the enzymes to digest the cellulose in wood, but bacteria and protozoa that live in their hindguts provide them with the necessary enzymes.

**Order Psocoptera (Psocids)**

The psocids, called booklice and barklice although they are not really lice, make up the order Psocoptera. Psocids are tiny, often no more than two tenths of an inch long. Most of them are winged. Some wingless species occur in buildings and are known as booklice because they eat the starch in the bindings of books. Most other species occur on bark, in clumps of dead leaves, or in other places where they can find the mold, lichens, or dead plant and insect tissue that they eat. A few species live in bird nests.

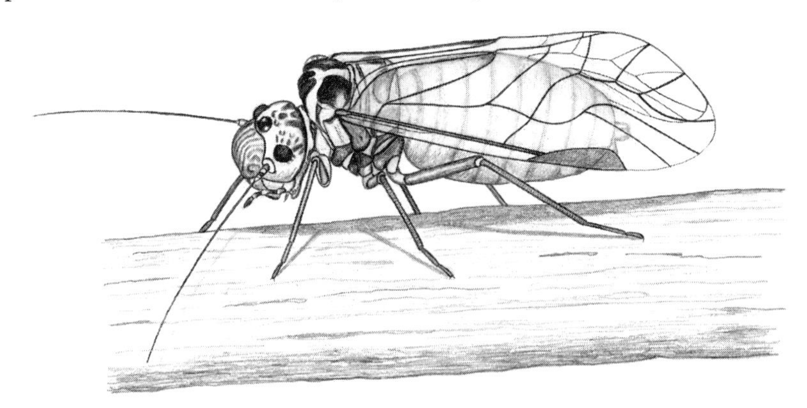

A psocid (actual length without antennae: 0.2 inch)

**Order Phthiraptera (Lice)**

The lice, the only members of the order Phthiraptera, are all small, flattened, and secondarily wingless, obvious adaptations to their parasitic life in the fur or hair of a mammal or among the feathers of a bird. They spend their entire lives on a host animal, mating there and gluing their eggs to its pelage or plumage. Transfers from one host to another are possible only when hosts are in close bodily contact, as during copulation or parental care. All sucking lice (once considered to constitute the separate order Anoplura) live on mammals and use their piercing-sucking mouthparts to drink blood. Most biting lice (once classified as the separate order Mallophaga) infest birds, but a few live on mammals. They use their chewing mouthparts to eat the feathers of birds and flakes of dry skin, bits of clotted blood, or other organic debris that they find on their hosts. A few use their mandibles to score the skin of the host to release blood, which they then lap up.

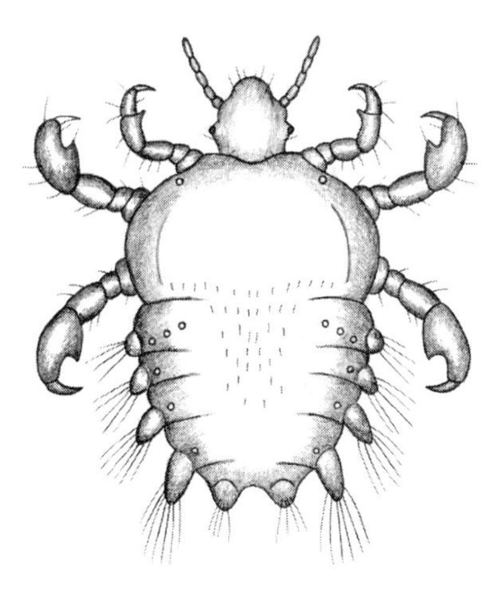

The crab or pubic louse of humans, a sucking louse (actual length: 0.08 inch)

**Order Hemiptera (True Bugs)**

The only insects that, very strictly speaking, are properly referred to as bugs are the members of the order Hemiptera. But even entomologists sometimes use "bugs" when they mean insects in general or particular non-hemipteran insects such as lightningbugs or ladybugs, which are actually beetles. All Hemiptera have piercing-sucking mouthparts. A few are wingless, but in the winged forms the hind wing is entirely membranous and the basal portion of the forewing is leathery, while the outer portion is membranous. The aquatic bugs, mostly predators, have not evolved gills but use a snorkel or come to the surface to get bubbles of air. The terrestrial bugs—the great majority of the order—are varied in habits and habitat.

Many suck the sap of plants; some suck the juices of other insects; and a few, such as the bed bugs, are blood feeders that attack birds or humans and other mammals.

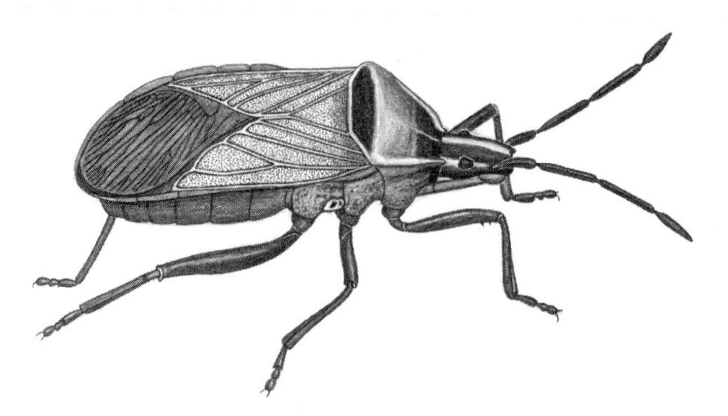

A true bug (actual length without antennae: 0.5 inch); note the piercing-sucking beak and the opening of a stink gland between the middle and hind legs

**Order Homoptera (Cicadas, Aphids, Leafhoppers, Scale Insects, and Others)**

There is no common name that includes all the insects in the order Homoptera: the cicadas, leafhoppers, planthoppers, whiteflies, aphids, scale insects, and others. They are closely related to the Hemiptera, and some authorities lump the two together as one order. The Homoptera have piercing-sucking mouthparts, and, without exception, feed on the sap of plants. Their front wings are homogeneous in texture, either all leathery, as in leafhoppers and planthoppers, or all membranous, as in cicadas and aphids. Some species of cicadas, the largest of the Homoptera, may be almost 2 inches long, but the rest of the Homoptera are much smaller, generally less than half an inch long. The periodical cicadas are notable because they spend many years underground sucking sap from the roots of trees. Some broods emerge en masse in the spring of their thirteenth year and others in the spring of their seventeenth year. Aphids are remarkable because of their unusual reproductive cycle. Most species overwinter as eggs in crevices in the bark of the twigs of woody plants. In the spring every one of the surviving eggs produces a wingless, parthenogenetic female, one that can reproduce without benefit of a male and that will give live birth to her young. All of her descendants for several generations will be parthenogenetically reproducing females that give live birth and are wingless except for a summer generation that migrates to a different species of plant, often an herbaceous one, and a late summer generation, consisting of both females and males, that migrates back to the original host plant and there reproduces sexually and lays the eggs that will survive the winter. Many of the scale insects lack eyes, antennae, and legs except as newly hatched nymphs and adult males. Throughout most of their lives, they are degenerate "parasites," bound to the plant on which they feed.

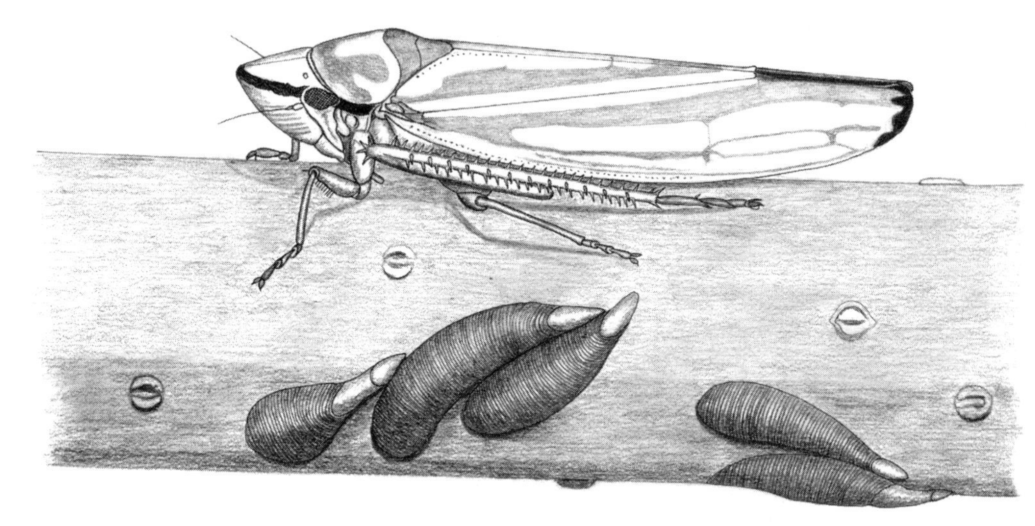

A leafhopper on a twig (actual length of leafhopper: 0.3 inch); below the leafhopper are several oyster shell scale insects covered by their waxy scales

## ORDERS WITH COMPLETE METAMORPHOSIS

### Order Neuroptera (Dobsonflies, Lacewings, Antlions, and Others)

Among the most familiar species of the order Neuroptera are the dobsonflies, the lacewings, and the antlions. The larvae of certain dobsonflies, known as hellgrammites when used as fish bait, have chewing mouthparts, are predators on other insects, and can be found under rocks in fast-flowing streams. Many common lacewings are green, even to the veins of the wings,

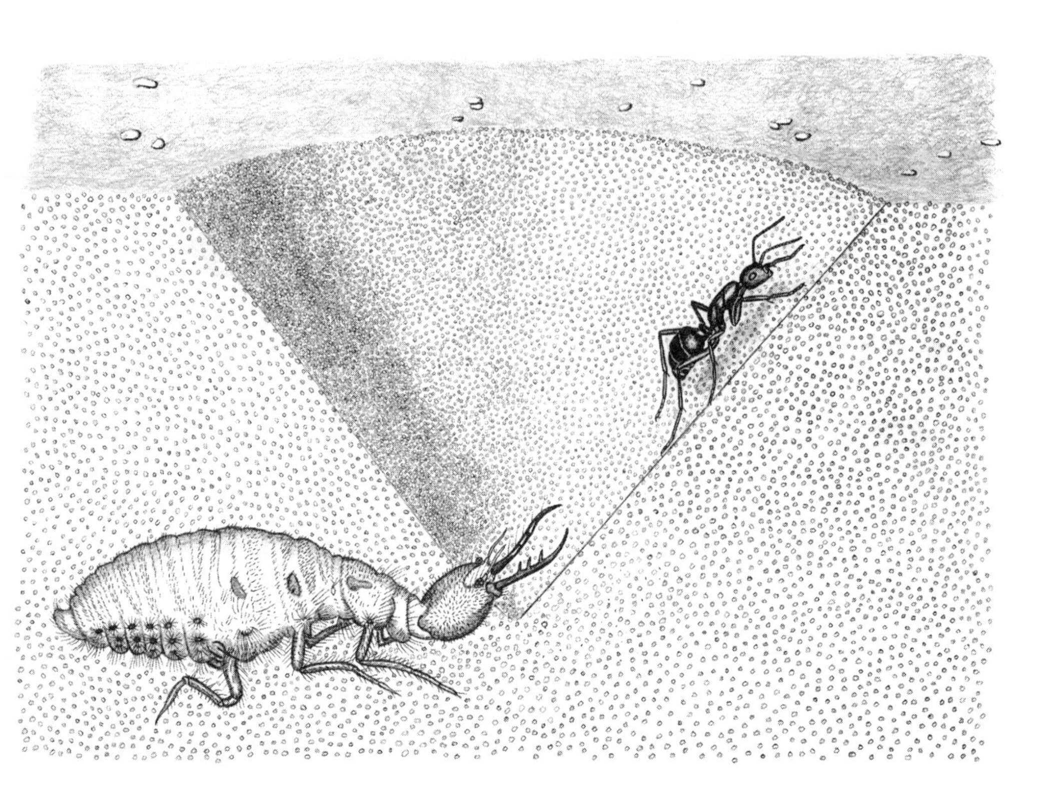

*Above:* An antlion larva and an ant trying to escape from the antlion's pit trap (actual length of antlion larva: 0.6 inch)

*Below:* An adult antlion (actual wingspan: 1.5 inches)

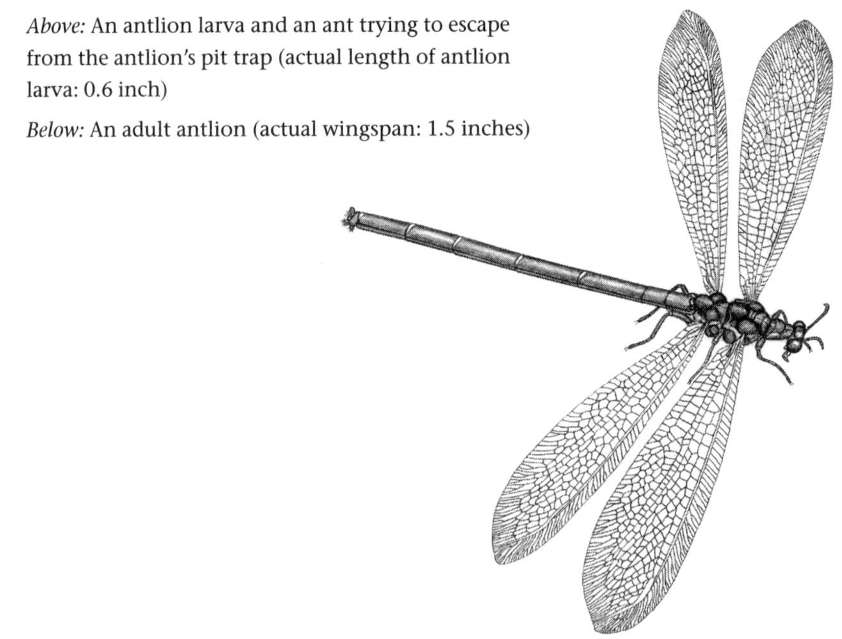

and have eyes that shine like burnished gold or copper. Adults use their chewing mouthparts to feed mainly on honeydew, especially if it is overgrown with sooty fungus. The eggs of lacewings are generally laid in loose clusters and are easily recognized because each egg is at the top of a tall, stiff stalk of silk. The larvae, called aphidlions, live in aphid colonies and suck the juice from their sluggish prey. Antlion larvae dig conical pits in dry, dusty soil and wait concealed at the bottom for an ant or some other small insect to stumble into their trap. The prey tumbles down the steep slope into the jaws of the antlion, often helped on by a bombardment of soil that the antlion flips upward with its flattened head.

**Order Coleoptera (Beetles and Weevils)**

The beetles and the weevils, the two large subgroups of the order Coleoptera, are so numerous, about 300,000 species, that they are difficult to summarize in capsule form. Some are aquatic, but most are terrestrial. Larval beetles vary in their eating habits. Some are predators on other insects; some eat plants; some are scavengers, sometimes in the nests of birds; only a few are parasites of other insects; and one species is a parasite that lives in the fur of beavers. The larvae have chewing mouthparts. Many have three

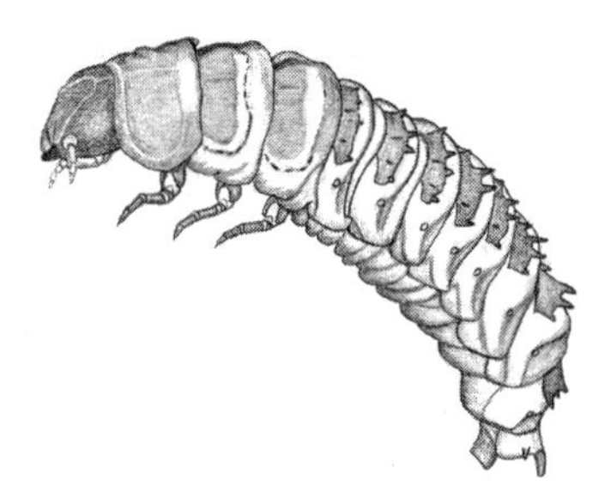

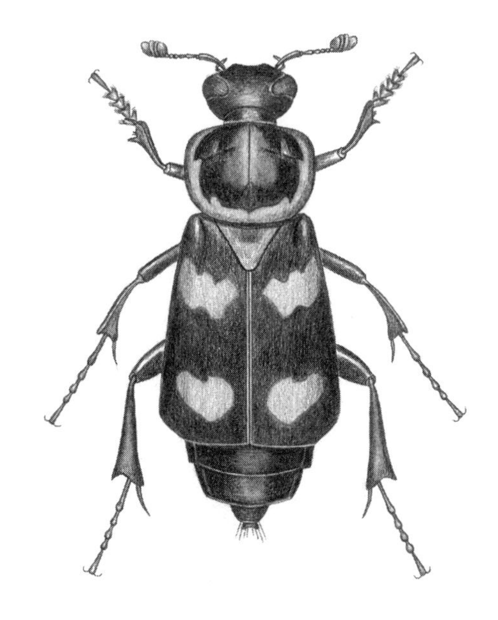

*Above:* A burying beetle larva (actual length 0.6 inch)

*Below:* An adult burying beetle (actual length: 0.7 inch)

pairs of thoracic legs, but others are legless. The adults also have chewing mouthparts, and are easily recognized by their elytra, horny, veinless front wings that have become a part of the body armor, protecting the abdomen and the membranous hind wings. Adults and larvae often eat the same food, but in some species they have different diets. Adult blister beetles, for example, eat foliage, while their larvae eat the eggs of grasshoppers or the eggs and the stored pollen and nectar of solitary bees. Adult burying beetles burrow back and forth under the carcass of a small dead animal, perhaps a mouse, until they have buried it. They and their larval offspring live beneath the carcass, where the parents care for their young by feeding them regurgitated carrion. Larvae of the notorious Japanese beetle live in the soil and eat the roots of grasses, while the adults eat the leaves of a variety of plants. Lightningbugs, or fireflies, which are really lampyrid beetles, eat other insects or snails in both the larval and adult stages. On June evenings, the adult males flash in the air as they signal to potential mates, which flash signals back from the foliage where they sit.

**Order Siphonaptera (Fleas)**

The order Siphonaptera consists solely of fleas. All the adults are small, laterally flattened, lack wings, have piercing-sucking mouthparts, and have their legs modified for leaping. Larvae have chewing mouthparts and slender, legless bodies that bear a few long bristles. The adults are usually to be found in the host's nest or sleeping place. They suck the blood of mammals or birds, leaping on and off the host as required. Fleas lay eggs in the host's nest or on its body. In the latter case the eggs soon fall to the host's sleeping place. There the larvae feed on organic debris, including dried droplets of blood that were defecated by adults.

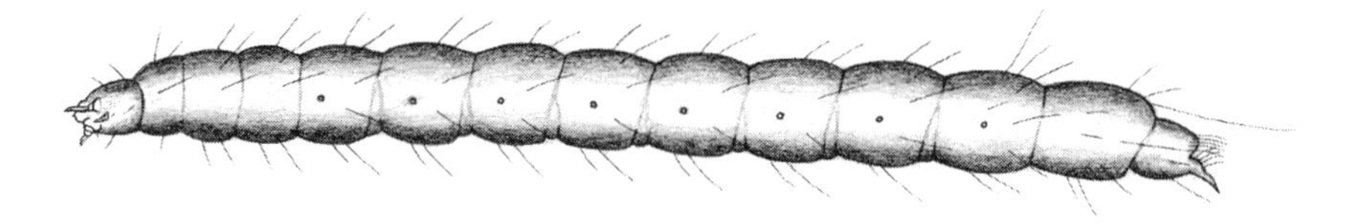

A flea larva (actual length: 0.15 inch)

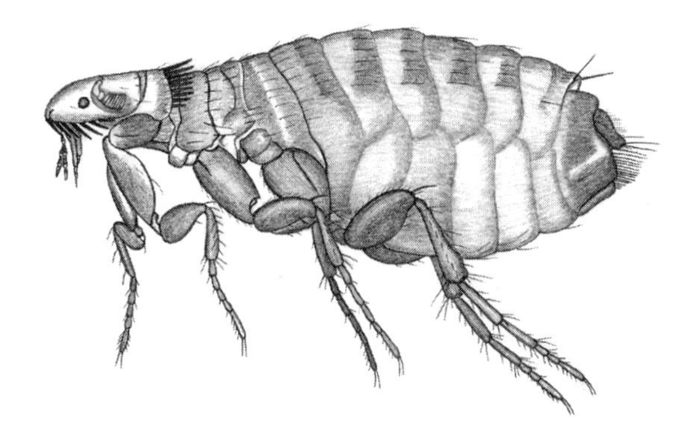

An adult flea (actual length : 0.1 inch)

**Order Diptera (Gnats, Mosquitoes, Flies, and Others)**

The order Diptera consists of the gnats, mosquitoes, flies, and others. The adults have only one pair of wings, the front ones. The hind wings have become tiny club-shaped organs, halteres, that beat rapidly as the insect flies and act as gyroscopes that sense when the insect deviates from stable flight by pitching, yawing, or rolling. While the more advanced Diptera, those that we usually refer to as flies, have short, three-segmented antennae, gnats, mosquitoes, and their primitive relatives have long, many-segmented antennae. The larvae of gnats, mosquitoes, and other primitive Diptera lack legs but have fully formed heads and mouthparts that are basically of the chewing type. By contrast, the larvae of the more advanced Diptera, called maggots, lack both legs and a fully formed head. Most of the head has been lost or withdrawn into the thorax, its only visible part a pair of small mouth hooks that protrude from the tapered end of the carrot-shaped body. The larvae of Diptera may be aquatic or terrestrial. They eat many things, including carrion, dung, rotting vegetation, other insects, and, in the case of mosquitoes and black flies, unicellular algae and other organic particles that they strain from the water in which they live. Some dipterous larvae are parasites of other insects or even of mammals or birds. Adults may feed on blood, nectar, or other substances—usually liquids—that they imbibe with their piercing-sucking or sponging mouthparts.

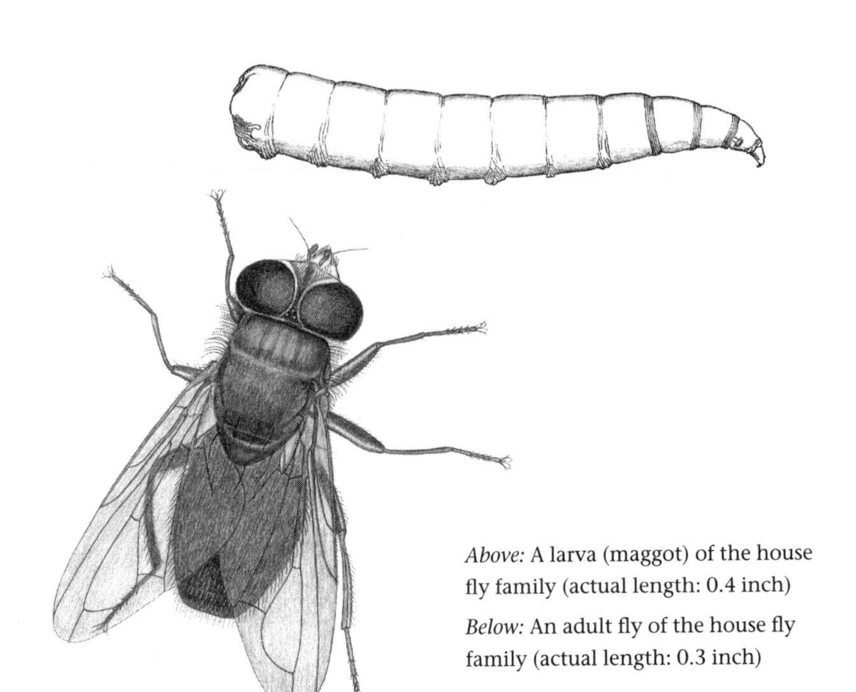

*Above:* A larva (maggot) of the house fly family (actual length: 0.4 inch)

*Below:* An adult fly of the house fly family (actual length: 0.3 inch)

**Order Lepidoptera (Moths, Skippers, and Butterflies)**

The order Lepidoptera contains the moths, skippers, and butterlies. Except for a few wingless females, all adults have large, broad, membranous wings that are made opaque by the minute scales that cover them like shingles on a roof. The antennae of moths are featherlike or threadlike and taper to a point. But all butterflies and skippers have threadlike antennae that end in a knob. In the skippers the knob ends in a hook. Most Lepidoptera have a long, thin, sucking proboscis, or tongue, that is coiled under the head like a watch spring when it is not in use. It is uncoiled and extended to suck nectar from flowers or to imbibe juices from carrion or rotting fruit. The larvae of the Lepidoptera, known as caterpillars, have long, cylindrical bodies with five or fewer pairs of short, stubby, fleshy prolegs on the abdomen; three pairs of short legs on the thorax; tiny antennae; and mouthparts of the chewing type. Most caterpillars eat leaves or other parts of plants, but a few are predators, and some, including the infamous clothes moth, eat our woolens and furs or the hair and dry skin that remain after other scavengers have eaten the soft tissues of a carcass.

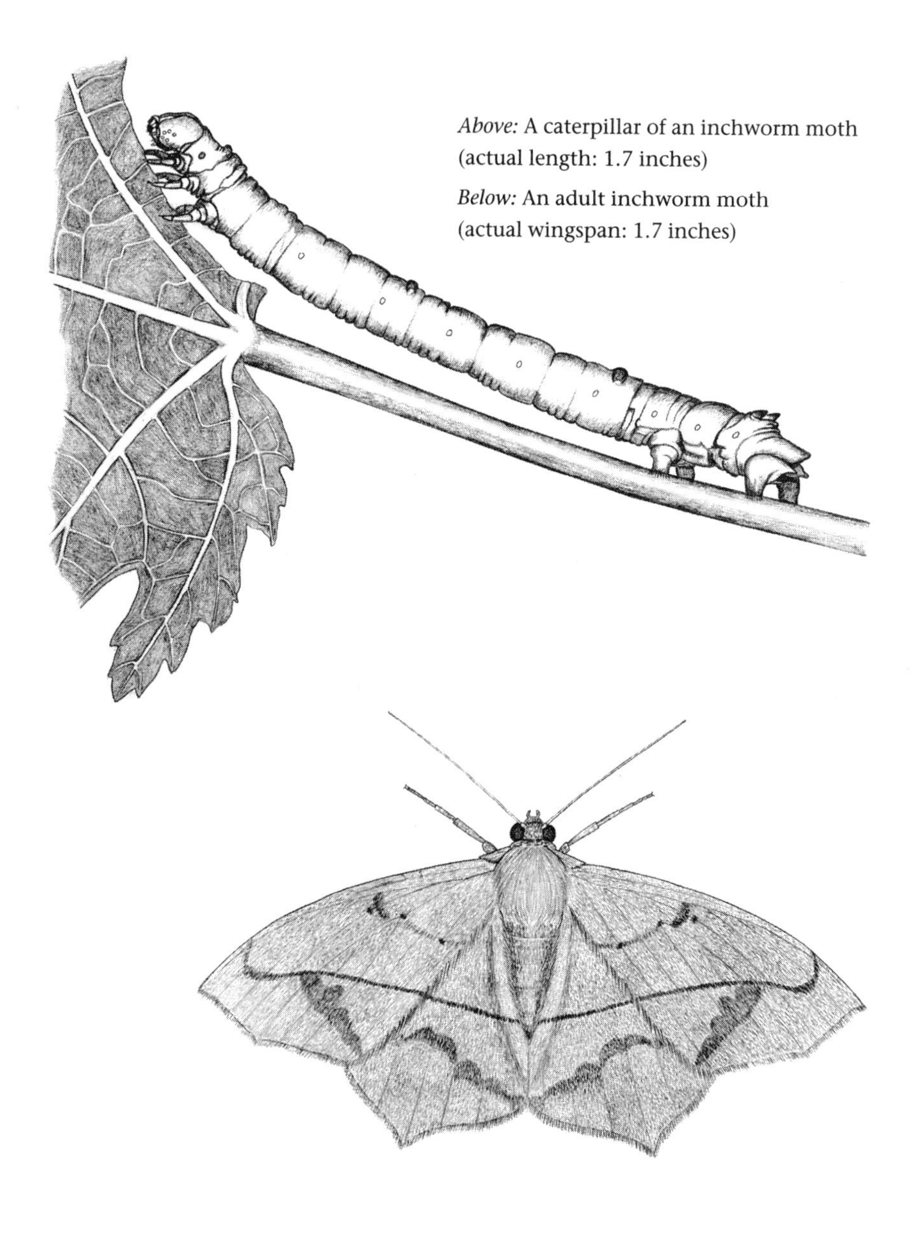

*Above:* A caterpillar of an inchworm moth (actual length: 1.7 inches)

*Below:* An adult inchworm moth (actual wingspan: 1.7 inches)

**Order Hymenoptera (Sawflies, Wasps, Ants, and Bees)**

The sawflies, parasitic wasps, predaceous wasps, ants, and bees constitute the order Hymenoptera. The mouthparts of adults are basically of the chewing type but in bees are modified for drinking nectar. The hind wings are much smaller than the front wings and are attached to them by a zipperlike structure. Except in the sawflies, named for their sawlike ovipositor, there is a narrow constriction between the apparent thorax and the apparent abdomen, technically known as the gaster. This constriction is actually behind the first segment of the abdomen, which is solidly joined to the hind part of the thorax. The ovipositor at the tip of the abdomen serves as a stinger in bees, many wasps, and some ants. Larval sawflies look much like caterpillars but have at least six pairs of abdominal prolegs rather than the five or fewer

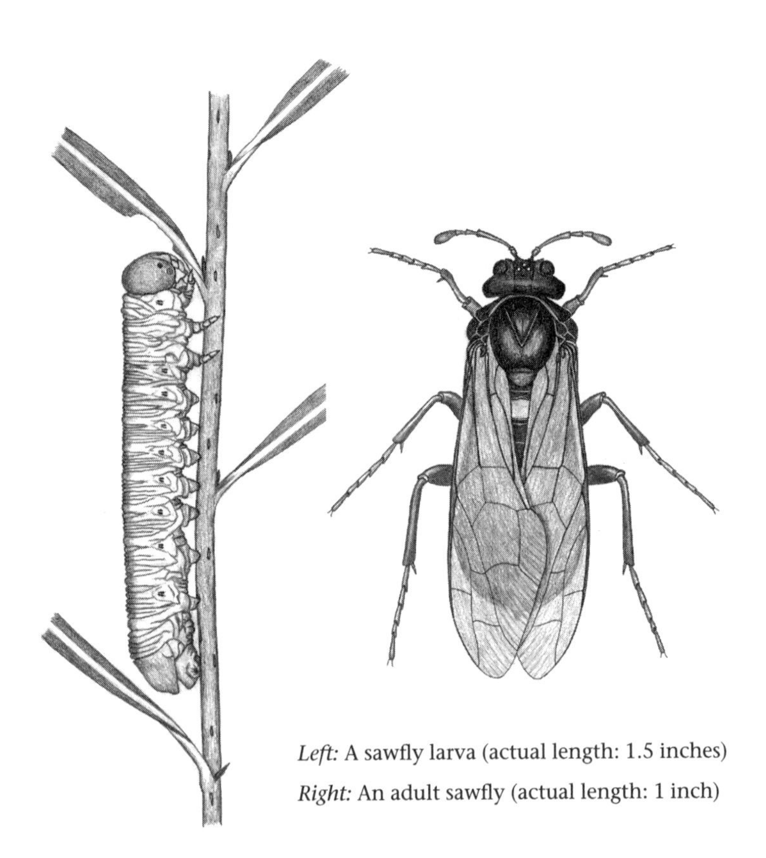

*Left:* A sawfly larva (actual length: 1.5 inches)
*Right:* An adult sawfly (actual length: 1 inch)

of caterpillars. The larvae of all of the other Hymenoptera are white, soft-skinned, and helpless, a condition they can tolerate because they are protected from many environmental dangers by being parasites within a host insect, by burrowing in a plant, by living within a gall, or by being cared for by their parents in a nest. Most social wasps feed their larvae insects. The social bees—the bumble bees and the honey bees—feed their larvae pollen and nectar, as do all other bees. The ants are all social and may be predators, herbivores, or omnivores.

What you have just read should enable you to identify to the level of the order about 98 percent of the insects that you see. But remember that most orders of insects and other animals are large and include a great diversity of species. The order of birds known as the Charadriiformes, for example, includes a worldwide total of 366 species that are classified in 19 different families that are comprised of such diverse forms as the jacanas, sandpipers, seedsnipes, thick-knees, avocets, pratincoles, plovers, gulls, terns, jaegers, auks, guillemots, and sandgrouse. Insect orders are even larger and are often even more diverse. The 300,000 species of beetles and weevils, the Coleoptera, are represented in America north of Mexico by about 30,000 species that are classified in 112 families that include plant feeders, predators, scavengers, and omnivores. Among them are insect-eating tiger beetles, predaceous diving beetles, aquatic whirligigs, rove beetles, carrion beetles, dung beetles, click beetles, hide beetles, ladybirds, leaf beetles, and weevils.

The determined amateur can learn to identify insects to the level of the family. There are about 613 families of insects in America north of Mexico. For this purpose, I recommend the field guide to the insects of America north of Mexico by Donald Borror and Richard White. But identification to the level of species is, with few exceptions, best left to the taxonomic experts. No one taxonomic entomologist is an expert on all groups of insects. These entomologists specialize, perhaps on one or more small orders, on a medium-sized order, or on one or a few related families of a large order such as the Coleoptera.

Butterflies have long been the province of amateur naturalists. They have collected them, many have become experts at identifying these beautiful

insects, and some have made noteworthy contributions to our scientific knowledge of butterflies. In recent years birders have come to consider butterflies, and more recently dragonflies, as "honorary birds," and now there is even a field guide that covers most of the coastal strip from Boston to Washington, D.C., *Butterflies through Binoculars,* by Jeffrey Glassberg, published in 1993. It gives field marks for identifying even the most obscure of the little brown butterflies of the region.

# Disappearing Diversity

# 10

At five different times during the last half billion years, numerous species of plants and animals disappeared more or less suddenly from the fossil record. The first of these mass extinctions, which occurred about 440 million years before the present, marked the end of the Ordovician, the geological period during which the first primitive vertebrates appeared. The second came about 365 million years ago at the end of the Devonian, the period that saw a great proliferation of fishes and the appearance of the first amphibians and insects. About 245 million years ago, near the end of the Permian, which is particularly notable for the proliferation of reptiles, there was a third great extinction. A fourth occurred about 210 million years before the present, marking the end of the Triassic, during which the first dinosaurs and mammals arose. The fifth and last of these extinctions, famous for the disappearance of the dinosaurs, took place about 65 million years ago at the end of the Cretaceous, the period that saw a great proliferation of insects and birds. Each one of these extinctions, according to E. O. Wilson, was followed by a long period of evolutionary repair that lasted 10 million years or more.

There is considerable geological evidence that the most recent of these extinctions, and probably the others too, were caused by an overwhelming physical catastrophe that affected the whole planet. Most scientists think that this disastrous event was the collision of a massive bolide, a meteorite or a comet, with the earth. The impact, as described by Wilson in *The Diversity of Life* (an indispensable book for anyone interested in the conservation of biodiversity), "kicked up an immense dust cloud that enshrouded the planet and then either cooled the atmosphere by blocking out the sun or else warmed it by trapping heat as in a greenhouse."

As Wilson has so compellingly asserted, we are now witnessing the early

stages of a sixth period of mass extinctions that may well exceed the others in magnitude, this one caused not by a physical catastrophe but rather by an ecological catastrophe, the actions of an exceptionally destructive member of the worldwide fauna. This extraordinarily destructive creature is, of course, the human being. Although the current period of extinctions is only beginning, humans have already exterminated thousands of species of plants and animals such as insects and birds through excessive exploitation, habitat destruction, and the introduction of exotic animals such as rats. But the extinctions that have already occurred are only a drop in the bucket compared with the flood of extinctions that we are causing right now and that will continue into the future at an ever accelerating rate if we do not soon take measures to conserve the plants and animals that are left. Biological diversity is threatened everywhere on earth.

The story of excessive predation by humans begins far back in prehistory. When the people that we now know as Native Americans arrived in North America from Siberia 11,000 to 12,000 years ago, they found a rich fauna of large mammals that had had no previous experience with human hunters. Among them were giant bisons, mammoths, mastodons, giant sloths, horses, and camels. These animals were no match for the newly arrived humans, such as the people of the Clovis culture with their sharp, fluted spear points. (I once found a Clovis point in a plowed field in central Illinois. It may have been used to kill a mammoth.) By 6,000 years ago most of these large mammals were gone. Many scientists think that they were hunted to extinction by the newly arrived humans.

Within a few hundred years of their arrival in New Zealand, about a thousand years ago, the Maoris, Polynesians who were the first human settlers of those islands, had hunted to extinction 13 or more species of moas, flightless birds that occurred only in New Zealand and that ranged in size from the size of a turkey to larger than an ostrich. To this day piles of moa bones mixed with the occasional artifact can be found on Maori hunting sites, most of which date from 700 to 900 years ago.

The ancient extinctions in North America and New Zealand are far from unique. Similar extinctions that coincided with the arrival of the first humans occurred in Madagascar and Australia—and also on numerous other islands of the Pacific, including Hawaii, as the Polynesians and their ancestors spread across the South Pacific in their outrigger canoes over a period of

A moa about 10 feet tall

several thousand years. They arrived in Samoa and other islands of central Polynesia as early as 2,000 B.C.E and reached their easternmost destination, Easter Island, as recently as 400 C.E.. The people of the outriggers, as Wilson wrote, "ate their way through the Polynesian fauna." Extinctions occurred on island after island. One of the islands of Tonga had 25 species of birds when the Polynesians arrived 3,000 years ago; only 8 of them survive today. As you already know, well over half of the 50 species of honeycreepers that survived on the Hawaiian Islands when Captain James Cook arrived in 1778 have since become extinct. But another 35 species, at the very least, had already been exterminated by the native Hawaiians by the time of Cook's arrival.

Extinctions due to excessive hunting continued throughout recorded his-

tory and are still going on today. If we consider only North American birds, the European settlers of the continent had exterminated 4 once abundant species by the end of the nineteenth century and had nearly exterminated 2 others. The flightless great auks, the original penguins, were safe when they fished far out on the North Atlantic, but when they nested in colonies on shore, people stole their eggs and clubbed them to death for their meat, oil, and feathers. As Peter Matthiessen has noted, the last 2, destined to become museum specimens, were killed on June 4, 1844. And the last authenticated specimen of a Labrador duck was killed by a gunner on Long Island on December 12, 1875. Flocks of Carolina parakeets once roamed our southeastern states, but over the years they were hunted down, mainly because they attacked orchard fruits. The last wild flock, 13 birds, was seen near Lake Okeechobee in Florida in April of 1904, and the last captive individual died in 1914. There was a time when great flocks of passenger pigeons obscured the sun, some flocks so large that they flew past for hours. Their nesting colonies occupied many square miles, and their nests were so numerous in the trees that heavy-laden boughs sometimes broke off and crashed to the ground. These birds were doggedly hunted for people's dinner tables and even for pig feed. The last really large nesting colony, near Petoskey, Michigan, in 1878, consisted of nearly a billion birds occupying an area 5 miles long and a mile wide. Market hunters wiped out the colony, shipping out five freight-car loads of pigeons every day for 30 days. After that, only relatively small nesting colonies were seen. The last commercial hunt for these rapidly disappearing birds took place in 1893, and the last passenger pigeon, a captive bird named Martha, died at the Cincinnati Zoological Gardens at 1:00 P.M. on September 1, 1914. As you will read below, Bachman's warbler and the ivory-billed woodpecker are probably extinct. Other birds have survived persecution by humans—but some of them just barely. Remnant populations of whooping cranes and Kirtland's warblers persist only because people are now willing to take extraordinary measures on their behalf. The Eskimo curlew is near extinction and only very rarely seen, a bird once so abundant that, like the passenger pigeon, it was shipped to market by the ton.

Many countries now have laws against the exploitation of at least some of their birds and other animals, sometimes even including certain insects (for example, Schaus's swallowtail, a large yellow and black butterfly of the Florida Keys that is threatened by collectors). Nevertheless, many creatures are still faced with extinction due to illegal collecting or hunting. Some birds, especially Central and South American parrots and macaws, are trapped for the trade in pet birds, a multimillion dollar industry that caters mainly to Americans. Worldwide, as N. J. Collar and A. T. Juniper have pointed out, at least 100 of the 330 species of parrots are endangered or threatened. In the New World alone, 42 species are near extinction, 22 of them solely or largely because they are being collected to be sold as pets. The little blue macaw (formerly known as Spix's macaw) has apparently been extirpated from its native habitat in Brazil and is now represented by only a few captive birds. The last known nest was plundered and the chicks offered for sale at the price of $40,000.

The tiger and the earth's five species of rhinoceros are all threatened with extinction by poachers who sell their body parts to Asians, especially the Chinese, who value them for their supposed medicinal properties. Some Chinese people, particularly impotent old men, believe that powdered rhinoceros horn is a powerful aphrodisiac, and as Robert Hendrickson wrote in *Lewd Food,* it is literally worth its weight in gold in some parts of Asia. Many parts of a tiger's body are thought to have various medicinal uses, but the penis is especially valued as an aphrodisiac. Remedies for impotence that actually work have recently become available through the efforts of Western medicine. Perhaps they will find their way to Asia and replace rhinoceros horn and tiger penis, thereby lessening the threat to these animals.

As far as I know, no bird has yet been driven to extinction by environmental pollution, including contamination by insecticides or other pesticides, but, as you will read below, there have been some close calls. No one knows how many insects and other invertebrate species have been exterminated by chemical pollution. Although no bird has been driven to extinction by pesticides, some populations of some birds were reduced to dangerously low levels by insecticides that contaminated the environment. In North America alone, as Paul Ehrlich and his coauthors have noted, the populations of several species, all of which feed at or near the top of a food chain, were reduced to distressingly low levels by chlorinated hydrocarbon insecticides, especially DDT. These insecticides, as you read in Chapter 7, are not readily biodegradable, are stored in the body fat of animals, and are, consequently, concentrated within food chains. An organism near the be-

ginning of a food chain, a plant or a plant-feeding insect for example, may accumulate a small and presumably harmless dose of 5 or 10 parts per million of an insecticide, but the insecticide becomes ever more concentrated as it is passed up the food chain, until the top predator at the upper end of the chain, such as the hawk that eats a songbird that ate many plant-feeding insects, accumulates an injurious dose of many hundreds of parts per million of the insecticide.

The peregrine, a bird-eating falcon, was actually extirpated from eastern North America by the use of chlorinated hydrocarbons, but that population has been reestablished by introductions from elsewhere. The newly established eastern population, as well as the other populations, has been growing steadily since the banning of DDT by the United States in 1972 and the subsequent banning of most of the other chlorinated hydrocarbon insecticides. The osprey, the bald eagle, and the brown pelican all eat mainly fish and are at or near the end of their food chains. They all capture live fish, but the bald eagle also scavenges dead fish when it can find them. Osprey and bald eagle populations were essentially unharmed in Alaska, where there is little agriculture and pesticides are seldom applied. But in the lower 48 states and much of southern Canada, populations of these birds were decimated by insecticides. And as you read in Chapter 5, brown pelicans were all but exterminated by insecticide pollution along the Gulf coasts of Louisiana and Texas, but are now recovering because rice and cotton farmers have switched to more environmentally friendly insecticides.

The intentional or unintentional introduction by humans of exotic organisms such as fungi, insects, or rats has wreaked havoc with ecosystems worldwide. American chestnuts, once among the most abundant and valuable trees of our eastern forests, were virtually exterminated—although an occasional sprout can still be seen—by the chestnut blight fungus, unwittingly brought to North America on nursery stock from Asia. Some Hawaiian honeycreepers were, as you read in chapter 5, wiped out by bird malaria after the unintentional introduction of a malaria-transmitting mosquito. Flightless, ground-nesting birds that evolved on islands with no mammalian predators, especially islands of the Indian and Pacific Oceans, were quickly driven to extinction by the depredations of foreign predators such as cats, rats, or mongooses. No North American bird has gone extinct due to the introduction of an exotic organism. But native cavity-nesting species such as eastern bluebirds, purple martins, and red-bellied woodpeckers face stiff competition for nesting sites from house sparrows and Eurasian starlings, intentionally introduced in 1852 and 1890, respectively.

Wanton exploitation, environmental pollution, and the introduction of exotic organisms continue to be serious threats to the existence of many species of animals, among them insects, fish, amphibians, reptiles, birds, and mammals. But many more animals, and many species of plants as well, are threatened with extinction by the destruction or alteration of their habitats.

Many North American birders who have been birding for a few decades will tell you that they now see far fewer warblers during the spring migration than they did back in the 1950s, 1960s, or even the 1970s. I have the same impression. In Connecticut in the 1940s, there were spring days on which great waves of migrants settled into the woodlands and warblers "dripped from the trees." I often saw many individuals of several species feeding in the same oak, among them nattily attired black-and-white warblers climbing about on the bark and butterfly-like redstarts and colorful palm, Nashville, yellow-rumped, Cape May, bay-breasted, and blackburnian warblers searching the unfurling leaves for caterpillars. In the 1950s and 1960s in central Illinois, I saw concentrations of warblers that were at least equally impressive—perhaps even more densely packed because they had to crowd into tiny areas of woodland surrounded by miles of crop fields. I have not seen anything comparable in many years. There is no doubt in my mind that the populations of most warblers have seriously declined during recent decades.

These warblers, as well as other migrants and even some nonmigratory birds, are declining in numbers largely because their habitats are being rapidly destroyed and fragmented—in the case of the migrants, not only their breeding habitats in North America but also their winter habitats in the Caribbean and Central and South America. Three of our North American birds, the ivory-billed woodpecker, Bachman's warbler, and the California condor, were driven to virtual extinction in recent decades, largely by the destruction of their habitats.

The ivory-billed woodpecker, which required a very large feeding territory in mature timber, was the victim of unrestrained logging. The last undoubted sighting of this bird in the United States occurred in 1946 in

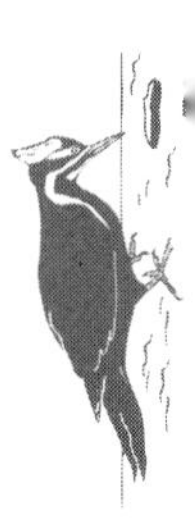

A male Bachman's warbler

Louisiana, but a few individuals may survive in the mountain forests of eastern Cuba.

Habitat loss due to the encroachment of agriculture and housing developments was largely responsible for eliminating the last free-living population of California condors from southern California, but pesticide residues and carrion laced with poison to control coyotes were also a factor. The last California condor known to exist in the wild was trapped on Easter Sunday of 1987 and joined 26 other condors in a captive breeding program at the San Diego and Los Angeles zoos. California condors reared in captivity are now being released in Arizona and California.

Bachman's warbler, which is almost certainly extinct, was the victim of the destruction of much of its breeding habitat in the bottomland swamps of the southern United States and the virtual elimination of its limited winter habitat in Cuba and the Isle of Pines, forests that have been almost totally cleared for the growing of sugar cane. According to H. David Bohlen, the last reliable sighting was made in Alabama in April of 1966. John James Audubon first published a description of this bird, naming it after the Reverend John Bachman, who had discovered it near Charleston, South Carolina, in 1833. Audubon's painting of this lovely warbler is prophetic—probably unintentionally so—in that he showed the bird perched on a blossoming branch of a tree that was extinct in nature even in Audubon's time. This tree, the mountainbay, now known as *Franklinia alatamaha* but known as *Gordonia pubescens* to Audubon, was discovered along the Alatamaha River near Barrington, Georgia, by John Bartram in 1765. It has not been seen in the wild since then, but trees descended from seeds and slips collected by Bartram are cultivated to this day. In 1974 Roger Tory Peterson wrote of the Bachman's warbler and *Franklinia:* "Time may prove the bird to be an even more irretrievable loss than the tree for plants can be perpetuated in cultivation, warblers cannot."

Several other North American birds are endangered or even threatened with extinction by the loss of their habitat. Among them is the marbled murrelet, a seabird of the Pacific coast that flies inland to nest in tall coniferous trees. This interesting bird is threatened by the logging of old-growth coniferous forests. The piping plover, which lays its eggs in a simple scrape on a sandy beach, is endangered because the increasing numbers of people and recreational vehicles on such beaches often crush the plovers' eggs. Everywhere in our southern states, the red-cockaded woodpecker is endangered by logging that destroys the old-growth forests that harbor the ancient, fungus-infected pines in which it excavates its nest cavities. The tiny California gnatcatcher, only recently recognized as a species distinct from the widespread black-tailed gnatcatcher, is threatened by the relentless urbanization that is eliminating its habitat, the creosote bush and mesquite scrub of southern California and Baja California. The nesting habitat of the endangered Kirtland's warbler, stands of young jack pines 6 to 18 feet high in a few counties in central lower Michigan, is disappearing because of human encroachment and human control of naturally occurring fires, which open the cones of jack pines so that they can shed their seeds. Other North American birds are also endangered or threatened by the loss of habitat: Florida scrub jays, golden-cheeked warblers and black-capped vireos in Texas, clapper rails in California, and the northern spotted owl in the Pacific Northwest.

Not only birds, but also plants, mollusks, insects, fish, amphibians, rep-

tiles, and mammals are threatened with extinction by habitat destruction—and not only in North America, but everywhere else on the earth, too. In the United States alone, for example, 57 species of clams, 22 snails, 29 insects, and 22 other arthropods are currently listed by the Division of Endangered Species of the U.S. Fish and Wildlife Service as endangered or imminently threatened with extinction. Hundreds, if not thousands, of other extinctions or potential extinctions have surely gone unnoticed, especially threats to such small and seldom noticed creatures as snails and insects. In Europe, the situation seems to be even worse than it is in the United States, but this may be only a matter of appearance. Things may only seem to be better in the United States, perhaps because we pay less attention to the problem or are more reluctant than Europeans to classify species as endangered. As E. O. Wilson has reported, 34 percent of 10,290 insects and other invertebrates were considered to be threatened or endangered in western Germany in 1987. In the same year 22 percent of 9,694 invertebrates were so classified in Austria, and in England 17 percent of 13,741 insect species.

But the most sweeping threat to biological diversity is the rapidly progressing destruction of tropical forests, especially the rain forest, the richest environment on earth, originally constituting less than 10 percent of the earth's land area but serving as the habitat of over half the earth's species of plants and animals. In prehistoric times there were about 5.5 million square miles of tropical rain forest. But by 1989 that had been reduced by somewhat more than half, to about 2.75 million square miles. This destruction continues at a rate of nearly 2 percent of the standing cover each year. As Wilson put it, "in 1989 the surviving rain forests occupied an area about that of the contiguous forty-eight states of the United States, and they were being reduced by an amount equivalent to the size of Florida every year." At this rate, these forests will be gone, except for a few preserves, by the year 2040, at about the time my grandson sends his children off to college.

Forests, including tropical forests—and along with them their animal inhabitants—are being destroyed for various reasons: to provide firewood, lumber, crop fields, pastures, and even space for urban expansion. Until recently, huge tracts of forest, mainly in Costa Rica, were being converted to cattle ranches that supplied cheap beef to chain hamburger restaurants in the United States and other developed countries. Adverse publicity and consumer boycotts ended this "hamburger connection," as Richard Primack termed it. Nevertheless, deforestation continues in Costa Rica and elsewhere, spurred by the need to provide wood, fiber, and food for a rapidly proliferating human population that grows largely unrestrained at a frightening rate. Just within my lifetime, the world population has almost tripled, soaring from about 2 billion to almost 6 billion. In *The Lost Gospel of the Earth,* Tom Hayden noted that the human population of the earth grows by about "90 million people yearly, eight new Calcuttas."

Ironically, land that once supported an exuberantly growing rain forest will usually be only marginally productive after it is cleared and planted with crops or pasture plants. When the forest grew there, the soil contained only very small quantities of the mineral nutrients required by plants. Most of these nutrients were held in the bodies of the trees, becoming available to new growth only when a tree fell and rotted. When the land was cleared and the trees were burned, the ashes temporarily charged the soil with nutrients. But in the peculiar soil of the tropical rain forest—very different from most soils in the temperate zone—these nutrients were soon rendered unavailable by chemical bonding or were leached deep into the soil, where they are out of reach of the crop plants. After two or three years, the land is likely to be abandoned because the soil has lost its fertility. It is different in the temperate zone. There forest soils are generally rich in plant nutrients and remain fertile long after the land is cleared.

A tract of rain forest will probably be lost forever once it has been cleared, almost certainly so if it does not border on a tract of intact rain forest. There are two principal reasons for this. First, the seeds of rain forest trees cannot remain dormant and survive for years as can the seeds of many temperate zone trees. The tropical seeds generally must germinate within a few days or weeks, long before animals or water currents can transport them to cleared land that is more than a short distance away. Second, even if a seed makes it to a cleared area and sprouts, the seedling will not survive if the soil has become too depleted of nutrients.

Many nations of the world, including Latin American countries such as Brazil, Costa Rica, and Peru, are setting aside small areas of forest and other dwindling habitats as nature preserves, as refuges for the thousands of species of plants and animals that are threatened with extinction by the destruction of their habitat. But the size of a preserve determines its usefulness

as a refuge. The question of how large preserves should be is being examined through an internationally funded research project in Brazil, originally known as the Minimum Critical Size of Ecosystems Project and now a part of the larger Dynamics of Forest Fragments Project. This research began in the late 1970s and is expected to continue into the next century.

It is well known that large islands can support more species than can small islands. And so it is with nature preserves—which are, after all, only fragments of a once more extensive habitat, islands in a sea of crop or pasture land. Even insects are more likely to do well in large areas than in small ones. Trapping with pitfall traps near Manaus in Brazilian Amazonia showed that dung beetles are much scarcer in clear-cut areas (717 individuals of only 12 species per plot) than they are in the intact forest (1,381 individuals of 44 species per plot of the same size). Forest fragments had more dung beetles than clear-cut areas but fewer than intact forest, both in numbers of species and in population densities. Low dung beetle populations resulted in an ecologically significant decrease in the rate at which dung decomposed, as Bert Klein has reported.

Creatures that live near the edge of the forest are more exposed to parasites and predators that invade from outside than are creatures that live deep within the forest. Detrimental edge effects are, as is to be expected, more severe in small refuges than in large ones—another reason why one large preserve in better than several small ones, even if the combined areas of the small preserves equal or even exceed the area of the large preserve. A small fragment has a greater proportion of edge than does a large one and its center is closer to the edge. Thus a small refuge will have a smaller, or even nonexistent, safe central area that is beyond the reach of predators or parasites that invade from the edge but will move only a limited distance inward. For example, a refuge that is 1,000 yards on a side has a large safe central haven that is not reached by a predator that will not go more than 100 yards from an edge. But an area that is only 200 yards on a side has no safe central haven that cannot be reached by this predator.

But information that has recently come to light shows that edges (also known as ecotones) are important factors in the evolutionary generation of biological diversity. A recent study in Africa directed by Thomas B. Smith, an evolutionary biologist at San Francisco State University, produced convincing evidence that ecotones, transition zones between adjacent habitats, are hotbeds of natural selection in which new species evolve far more often than they do in the adjoining "pure" habitats. Thus, as Smith found, the zone of transition between African rain forest and savannah, where clumps of trees intermingle with grassland, may be the major generator of rain forest biodiversity, the nursery that produced, and doubtlessly continues to produce, new species that ultimately shift to the rain forest habitat. Much the same thing must be happening in ecotones everywhere, including those between the rain forest and other habitats in Central and South America. Conservationists have so far paid little attention to the preservation of these transition zones, but it is to be hoped that this will change. As quoted in a news release in *Science* in 1997, Smith said, "If we are to protect biodiversity, we should also protect the processes that generate it."

The ever increasing impact of brown-headed cowbirds on North American songbird populations is an ecologically significant and well-documented example of a detrimental edge effect. This bird is a brood parasite that lays its eggs in the nests of other birds, mainly migrant songbirds, which care for the nestling cowbirds to the disadvantage of their own young, which rarely survive to leave the nest.

The introduction of European agriculture to North America created vast new areas of open habitat that are suitable for brown-headed cowbirds, and these birds now occupy all of the continental United States and southern Canada from coast to coast. Arthur Cleveland Bent wrote that this cowbird is "supposed. . . to have entered North America through Mexico, to have spread through the Central Prairies and Plains with the roving herds of wild cattle (bison), and to have gradually extended its range eastward and westward to the Atlantic and Pacific coasts as the forests disappeared and domestic cattle were introduced on suitable grazing lands."

The spread of this brood parasite has had a detrimental effect on many songbirds, and as the remaining woodland remnants become ever more fragmented, the plight of migrants from the American tropics, such as thrushes, vireos, and warblers, becomes ever more serious. Although brown-headed cowbirds are not woodland birds, they do penetrate woodlands for some distance beyond the edge to parasitize the birds that nest there. A small fragment of woodland is "all edge," and there is no longer a safe haven in a deep interior where cowbirds do not penetrate.

Several studies, summarized by Scott Robinson of the Illinois Natural

History Survey, show clearly that the rate of cowbird parasitism on neotropic migrants nesting in midwestern oak-hickory forests increases with the fragmentation of the forests, with the prevalence and proximity of edges that abut on open country. In small woodlots (of less than 250 acres) isolated in a sea of corn and soybeans in central Illinois, 75 percent of the nests of migrant songbirds had been parasitized. In the Shawnee National Forest in southern Illinois, which has many scattered open areas but has 40 to 60 percent forest cover, only 50 to 60 percent of the nests had been parasitized. Less than 30 percent of the songbird nests in the Hoosier National Forest in southern Indiana, which is more than 80 percent forested, contained cowbird eggs. In the central Missouri Ozarks, more than 90 percent covered by forest, less than 5 percent of the nests of migrant songbirds had been parasitized by cowbirds. Finally, R. T. Holmes and his colleagues, although not working in the Midwest, never found a parasitized nest in the Hubbard Brook Experimental Forest in New Hampshire, which is virtually entirely covered by forest.

There is no doubt that cowbirds do decrease populations of migrant songbirds, all of which are at least partly insectivorous. But what, if any, are the broad ecological implications of these declines? For example, do songbirds significantly decrease insect populations? If so, what impact does this natural control of insects, including many plant-feeding species, have on the plants or other organisms of the forest?

As you read in Chapter 3, birds have evolved a multitude of ways, some highly specialized, to exploit insects as food—many more ways than I can recount in this book. Birds beset insects from all sides, and virtually no species of insect is safe from their attack. Victor Shelford, often recognized as the father of modern animal ecology, estimated that during the 60 days of early summer, when birds are breeding, a square mile of deciduous forest in the eastern United States is populated by almost 2.7 billion invertebrates, mostly insects and spiders, and about 768 pairs of small nesting birds. He calculated that these birds and their young eat about 38.6 million insects during this period, over 500 per nesting pair per day. These are impressive numbers, but they do not answer the question of how important the impact of birds upon insects is to maintaining the "balance of nature." In other words, these numbers do not prove that birds eat enough insects to exert a significant controlling effect on the growth of insect populations.

Most insects lay hundreds of eggs. If only two of these eggs survive to become reproducing adults, the population size will be the same in the next generation. If four survive, the population will double. Thus hundreds of eggs, larvae, pupae, or adults that have not yet become parents must be eaten or otherwise eliminated if a population is to remain stable. Until recently, most ecologists thought that birds do not exert a significant controlling effect on insect populations. But as you will read below, recent experiments done in a western conifer forest and in a midwestern hardwood forest indicate that birds really are important in suppressing insect populations.

The economic ornithologists of the late nineteenth and early twentieth centuries believed that their analyses of the stomach contents of birds, which show that they eat many insects, proved that birds are a major factor in the suppression of insect populations. But by the 1930s ecologists and other biologists had realized that stomach contents alone are not sufficient evidence. Such data say nothing about what proportion of an insect population birds destroy. Until recently, possibly as a reaction to the unfounded claims of the early economic ornithologists, most ecologists assumed that birds are of little importance in limiting the growth of animal populations, including insect populations. They argued that the population growth of animals is limited mainly by environmental factors such as weather and by the amount of food available in the preceding level of the pyramid of biomass rather than by predation by animals in the succeeding level of the pyramid. But recent experiments indicate that these ecologists are probably wrong, and that the early economic ornithologists were probably right, although their view was essentially a "lucky guess" not supported by relevant data.

In the 1970s Torolf Torgersen and his colleagues at the Forestry and Range Sciences Laboratory in La Grande, Oregon, did an experiment that had been begging to be done, an experiment that addressed the question of just how much control birds actually exert on an insect population. They measured the impact of insect-eating birds on western spruce budworms, caterpillars that, during outbreaks that occur at intervals of 20 years or more, are so numerous that they kill Douglas firs and other conifers outright by stripping them of all their foliage. Torgersen and his colleagues enclosed Douglas firs that were as much as 30 feet tall in huge cages covered with

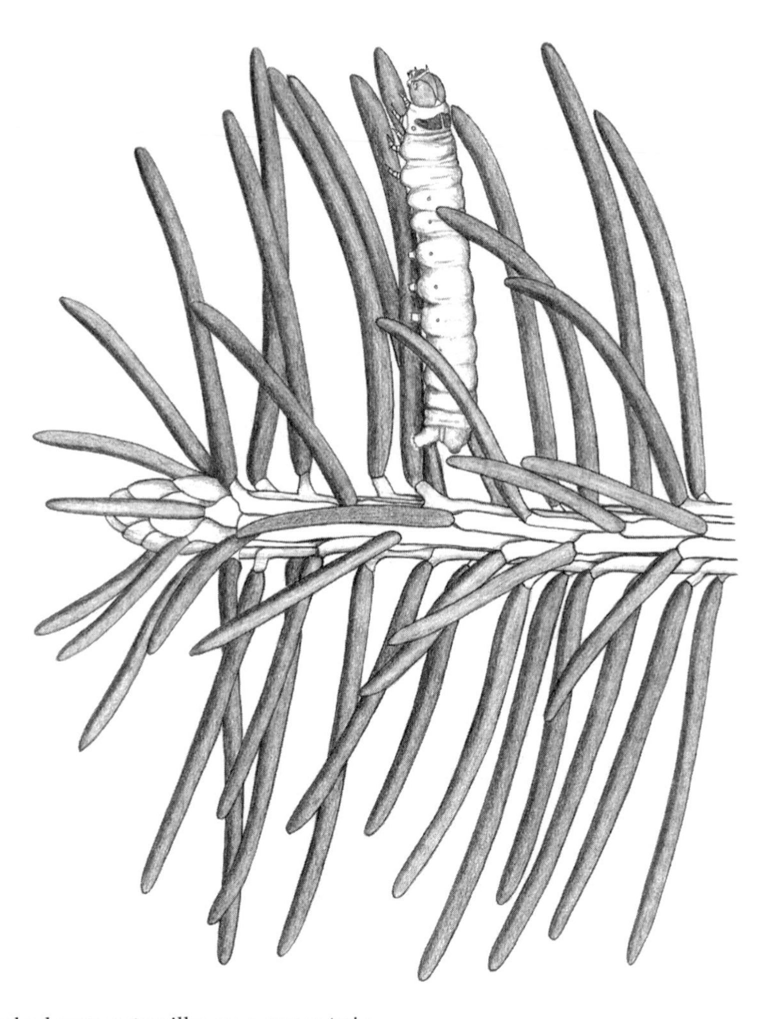

A spruce budworm caterpillar on a spruce twig

mesh that was too fine to admit birds but coarse enough to let insects pass. (By observing uncaged trees they had already found that birds eat large numbers of budworm caterpillars—birds of at least 27 species, among them evening grosbeaks, Cassin's finches, western tanagers, mountain chickadees, and golden-crowned kinglets.)

Comparing the numbers of spruce budworms that survived to the adult stage on caged and uncaged trees showed that birds destroy a great many caterpillars of this species. Caged trees, not accessible to birds, produced about six times as many budworm moths as did uncaged trees that were accessible to birds and on which birds had actually been seen to eat budworm caterpillars.

Robert Campbell, Torgersen, and Nilima Srivastara found that predaceous ants that forage in the foliage of conifers are also important predators of spruce budworms. Budworm survival increased when ants were excluded from caged trees by the application of a wide band of sticky material to the trunk of the tree just below the beginning of the crown. Trees from which both birds and ants were excluded produced up to 15 times as many budworm moths as did trees to which birds and ants had free access.

Robert Marquis of the University of Missouri at St. Louis and Christopher Whelan of the Morton Arboretum in Lisle, Illinois, did a similar experiment but broadened its ecological scope by comparing the impact on the trees of uncontrolled and bird-controlled insect populations. In 1989 they covered groups of white oak saplings with nets with a mesh too fine to admit birds but coarse enough to admit insects. Control groups of saplings were left uncovered. Comparing these two groups with a third group that was uncovered but sprayed with an insecticide, revealed a cascade of effects that went from the top of the food chain down to its base. Birds reduced the population of plant-feeding insects by half, which decreased by half the leaf area lost to chewing insects, which, in turn, made it possible for the aboveground growth of the saplings to increase by a third. There is little doubt that insectivorous birds do limit the growth of insect populations, and thereby significantly enhance the growth of trees and probably other plants. It seems clear that birds are indispensable members of terrestrial ecosystems.

The fragmentation of a habitat is likely to have wide-ranging ecological repercussions, as may the extinction of a species, the decline of a population, or any other environmental perturbation. The fragmentation of midwestern woodlands, for example, has had many harmful effects, among them a concatenation of events stemming from the increased access of cowbirds to the nests of songbirds, to nests that they could not have reached in an intact forest. As a consequence, songbird populations have declined and since there are fewer birds, they eat fewer insects that destroy the foliage and other parts of the woodland trees. These increasing insect

populations do more damage to trees and thereby slow their growth, especially the growth of saplings. Slower growth impairs the woodland's capacity to recover from the effects of severe storms or other cataclysms that knock down or otherwise destroy mature trees. The forest fragment thus becomes an even less suitable haven for the birds and other creatures that are its natural inhabitants.

⁎ The essence of what I have so far said in this chapter is that the great majority of current extinctions are caused by the activities of people, ranging from the wanton exploitation of one or a few species to the destruction of habitats that are home to thousands of species. Another plant or animal species disappears forever from the earth about once every 15 minutes. Even as I write these lines, somewhere a lonely creature, the last of its kind, is drawing its final breaths. This last survivor may be a songbird, perhaps a male Bachman's warbler that has gone unnoticed by birders, a bird that sang to attract a female that never came, a bird that was ready to defend a territory against an intruder that never came, a bird whose territory would never see a new generation of Bachman's warblers.

The ornithologist H. David Bohlen of the Illinois State Museum in Springfield dedicated his recent book, *The Birds of Illinois,* to the birders of the state, "who search for remnants of natural beauty in a fragmented and impoverished landscape." Impoverished is an understatement. Much of Illinois was originally tall-grass, black-soil prairie interspersed with an occasional grove of hardwood trees. The groves are largely gone, represented now only by small, scattered woodlots, and the prairie has virtually vanished. Originally there were 13 million acres of this prairie in eastern Illinois and northwestern Indiana, but as of 1985 only a few fragments totaling 20 acres remained undisturbed; the rest has been plowed and converted to farmland. As you read above, the tropical rain forests face the same fate. Assuming that their destruction continues at the present rate, they will, except for a few protected remnants, be gone by the year 2040. Will these remnants preserve all or most of the species of the rain forests? If not, and this may be what actually happens, tens of thousands of species will disappear from the earth along with the rain forests.

Natural selection has been kind to humans, molding us to be the dominant species on the planet—"the lords o' the creation," as Robert Burns wrote. Our versatile anatomy, our immense mental capacity, and our cooperative social nature give us enormous and menacing power over the other creatures and plants of our world. We have the capacity to destroy or to preserve. So far, from an ecological perspective, we have been mainly destroyers, preserving but little as, for the sake of progress and profit, we unthinkingly wipe out whole habitats and exterminate species by the thousands. In so doing we both threaten our ultimate material well-being and diminish our spirit.

The purely material reasons for adopting a "conservation ethic" and preserving biological diversity are too numerous to list here in their entirety. The potential long-term benefits are immense. Controlling the runaway growth of the human population and adopting ecologically sound agricultural practices would, for example, halt the desertification of semi-arid areas such as the savannahs of the Sahel of northern Africa. In this area, overly intensive farming and grazing, responses to unrestrained population growth, are converting a habitat that can support moderate human populations to a desert that can support virtually no one. An Amazonian rain forest is worth more, from a purely financial aspect, in its virgin state than it is after it has been cleared and put to some other use that people imagine to be profitable. Charles Peters and two colleagues reported that a hectare (2.4 acres) of rain forest in Peru would yield a one-time net income of $1,000 if it is clear cut for lumber. But it will yield, into the indefinite future, a yearly net income of $422 if its wild fruits, nuts, and rubber are harvested. On this basis and in the language of economists, the *net present value* of the virgin hectare is $6,330, calculated from anticipated returns discounted for the interest that could be earned if the sale price were otherwise invested. This is far more than what the *net present value* (based on gross returns) would be if the same hectare were to be put to some more conventinal use—$3,184 if it is cleared and converted to a pulpwood plantation, $2,690 if it is cleared and converted to pasture.

But the most important practical reason for preserving biological diversity is that the animals and especially the plants of the world are a virtually untapped source of as yet undiscovered riches: potential new crop plants, genes for improving current agricultural crops, heretofore unknown parasites and predators that might be used as biological agents to control pestif-

erous insects or plants, and new chemical compounds that could be used as medicines. For example, vincristine and vinblastine, recently discovered in a periwinkle plant from Madagascar, are potent drugs for the treatment of blood cancers. Their use has increased the survival rate of victims of childhood leukemia from about 10 percent to about 90 percent. To date, the plants and animals of the world have been the major sources of medicines. Aspirin, quinine, antibiotics, many antitumor agents, and a host of other drugs are derived from plants or animals or are synthetic copies or variants of naturally occurring substances. We don't know how many more valuable drugs remain to be discovered, drugs that will be lost to us forever if the plants or animals that contain them are driven to extinction.

The material benefits of preserving habitats and diverse floras and faunas are obviously immense and ultimately incalculable, but far more important are the less obvious and often ignored benefits to the human spirit. Some people seem to care very little about the preservation of humanity's natural heritage: "Let's cash in on that old-growth forest. We'll clear cut and sell the logs to Japan." But other people do care very much. In *The Primordial Bond,* Stephen Schneider and Lynn Morton quote Roman Vishniac: "All life is interrelated. To understand man, one must understand all living things." Every living species has something to tell us about life and about ourselves. William Blake saw "a world in a grain of sand and a heaven in a wild flower." E. O. Wilson stated it beautifully in *Biophilia:*

> Now to the heart of wonder. Because species diversity was created prior to humanity, and because we evolved within it, we have never fathomed its limits. As a consequence, the living world is the natural domain of the most restless and paradoxical part of the human spirit. Our sense of wonder grows exponentially: the greater the knowledge, the deeper the mystery and the more we seek knowledge to create new mystery.

The greatest of all human adventures is the quest for self-knowledge. But since we are but one aspect of the planetary web of life, we can understand ourselves only in the context of all life. The loss of a species or the destruction of a habitat narrows our view of ourselves, thus diminishing us and impoverishing our spirit.

# Selected Readings

## 1. Bugs and Birds through the Ages


Bodenheimer, F. S. 1951. *Insects as Human Food.* The Hague: W. Junk.

Borror, D. J., G. A. Triplehorn, and N. F. Johnson. 1989. *An Introduction to the Study of Insects.* Philadelphia: Saunders College Publishing.

Borror, D. J., and R. E. White. 1970. *A Field Guide to the Insects of America North of Mexico.* Boston: Houghton Mifflin.

Diamond, J. M. 1966. Zoological classification of a primitive people. *Science,* 151:1102–1104.

Elzinga, R. J. 1987. *Fundamentals of Entomology,* 3rd. ed. Englewood Cliffs, N.J.: Prentice-Hall.

Evans, H. E. 1984. *Insect Biology.* Reading, Mass.: Addison-Wesley.

Feduccia, A. 1996. *The Origin and Evolution of Birds.* New Haven: Yale University Press.

Gill, F. B. 1994. *Ornithology,* 2nd. ed. New York: W. H. Freeman.

Ross, H. H., C. A. Ross, and J. R. P. Ross. 1982. *A Textbook of Entomology.* New York: John Wiley and Sons.

Turbott, E. G., ed. 1967. *Buller's Birds of New Zealand.* Christchurch, New Zealand: Whitcombe and Tombs.

Waldbauer, G. 1996. *Insects through the Seasons.* Cambridge, Mass.: Harvard University Press.


## 2. The Only Flying Invertebrates


Borror, D. J., C. A. Triplehorn, and N. F. Johnson. 1989. *An Introduction to the Study of Insects.* Philadelphia: Saunders College Publishing.

Bray, O. E., J. L. Kennely, and J. L. Guarino. 1975. Fertility of eggs produced on territories of vasectomized red-winged blackbirds. *The Wilson Bulletin,* 37:187–195.

Briskie, J. V. 1992. Copulation patterns and sperm competition in the polygynandrous Smith's Longspur. *Auk,* 109:563–575.

Brower, L. P., B. S. Alpert, and S. C. Glazier. 1970. Observational learning in the feeding behavior of Blue Jays (*Cyanocitta cristata* Oberholser, Fam. Corvidae). *American Zoologist,* 10:475–476.

Davies, N. B. 1983. Polyandry, cloaca-pecking, and sperm competition in dunnocks. *Nature,* 302:334-336.

———1992. *Dunnock Behavior and Social Evolution.* New York: Oxford University Press.

Dethier, V. G. 1992. *Crickets and Katydids, Concerts and Solos.* Cambridge, Mass.: Harvard University Press.

Evans, H. E. 1984. *Insect Biology.* Reading, Mass.: Addison-Wesley.

Gibbs, H .L., P. J. Weatherhead, P. T. Boag, B. N. White, L. M. Tabak, and D. J. Hoysak. 1990. Realized reproductive success of polygynous red-winged blackbirds revealed by DNA markers. *Science,* 250:1394–1397.

Gill, F. B. 1994. *Ornithology,* 2nd. ed. New York: W. H. Freeman.

Linsenmaier, W. 1972. *Insects of the World.* Trans. L. E. Chadwick. New York: McGraw-Hill.

Place, A. R., and E. W. Stiles. 1992. Living off the wax of the land: bayberries and yellow-rumped warblers. *Auk,* 109:334–345.

Shugart, G. W. 1988. Uterovaginal sperm-storage glands in sixteen species with comments on morphological differences. *Auk,* 105:379–385.

Stiles, F. G. 1975. Ecology, flowering phenology, and hummingbird pollination of some Costa Rican *Heliconia* species. *Ecology,* 56:285–301.

Storer, R. W. 1971. Adaptive radiation of birds. In D. S. Farner, J. R. King, and K. C. Parker, eds., *Avian Biology,* vol. 1. New York: Academic Press.

Tomback, D. F. 1980. How nutcrackers find their seed stores. *Condor,* 82:10–19.

Vander Wall, S. B., and H. E. Hutchins. Dependence of Clark's nutcracker, *Nucifraga columbiana,* on conifer seeds during the postfledging period. *Canadian Field Naturalist,* 97:208–214.

Waldbauer, G. 1996. *Insects through the Seasons.* Cambridge, Mass.: Harvard University Press.

Wigglesworth, V. B. 1945. Transpiration through the cuticle of insects. *Journal of Experimental Biology,* 21:97–113.

Willson, M. F. 1983. *Plant Reproductive Ecology.* New York: John Wiley and Sons.

## 3. Bugs That Birds Eat

Beal, F. E. L. 1900. *Food of the Bobolink, Blackbirds, and Grackles.* U.S. Department of Agriculture Biological Survey, Bulletin 13.

Bent, A. C. 1939. *Life Histories of North American Woodpeckers.* U.S. National Museum, Bulletin 174.

———1942. *Life Histories of North American Flycatchers, Larks, Swallows, and their Allies.* U.S. National Museum, Bulletin 179.

———1949. *Life histories of North American Thrushes, Kinglets, and their Allies.* U.S. National Museum, Bulletin 196.

———1958. *Life Histories of North American Blackbirds, Orioles, Tanagers, and Allies.* U.S. National Museum, Bulletin 211.

Brandt, H. 1951. *Arizona and its Bird Life.* Cleveland: The Bird Research Foundation.

Brower, L. P. 1969. Ecological chemistry. *Scientific American,* 220:22–30.

Brower, L. P., and W. H. Calvert. 1985. Foraging dynamics of bird predators on overwintering monarch butterflies in Mexico. *Evolution,* 39:852–868.

Bryens, O. M. 1926. Actions of the northern pileated woodpecker. *Auk,* 43:98.

Campbell, B., and E. Lack, eds. 1985. *A Dictionary of Birds.* Vermillion, S.D.: Buteo Books.

Chapman, F. M. 1907. *The Warblers of North America.* New York: D. Appleton.

Christian, K. A. 1980. Cleaning/feeding symbiosis between birds and reptiles of the Galapagos Islands: new observations of inter-island variability. *Auk,* 97:887–889.

Craighead, F. C. 1950. *Insect Enemies of Eastern Forests.* U.S. Department of Agriculture, Miscellaneous Publication 657.

Diamond, A. W., and A. R. Place. 1988. Wax digestion in black-throated honeyguides. *Ibis,* 130:558–561.

Felt, E. P. 1905. *Insects Affecting Park and Woodland Trees.* Albany, N.Y.: New York State Museum.

Fink, L. S., and L. P. Brower. 1981. Birds can overcome the cardenolide defence of monarch butterflies in Mexico. *Nature,* 291:67–70.

Forbush, E. H. 1907. *Useful Birds and Their Protection.* Boston: Massachusetts State Board of Agriculture.

———1927. *Birds of Massachusetts,* vol. 2. Massachusetts Department of Agriculture.

Friedman, H. 1929. *The Cowbirds.* Springfield, Ill.: Charles C. Thomas.

———1955. *The Honey-guides.* U.S. National Museum, Bulletin 208.

Glick, P.A. 1939. *The Distribution of Insects, Spiders, and Mites in the Air.* U.S. Department of Agriculture, Technical Bulletin 673.

Greenstone, M. H., R. R. Eaton, and C. E. Morgan. 1991. Sampling aerially dispersing arthropods: a high-volume, inexpensive, automobile-aircraft-borne system. *Journal of Economic Entomology,* 84:1717–1724.

Hamilton, W. J., Jr. 1951. The food of nestling bronzed grackles, *Quiscalus quisculas versicolor,* in central New York. *Auk,* 68:213–217.

Heatwole, H. 1965. Some aspects of the association of cattle egrets with cattle. *Animal Behaviour,* 13:79–83.

Heinrich, B., and S. L. Collins. 1983. Caterpillar leaf damage and the game of hide-and-seek with birds. *Ecology,* 64:592–602.

Henderson, J. 1927. *The Practical Value of Birds.* New York: MacMillan.

Irwin, T. L. 1983. Tropical forest canopies: the last biotic frontier. *Bulletin of the Entomological Society of America,* Spring 1983:14–19.

Jackson, T. H. E. 1945. Some *Merops-Ardeotis* perching associations in northern Keyna. *Ibis,* 87:284–286.

Johnson, R. A. 1954. The behavior of birds attending army ant raids on Barro Colorado Island, Panama Canal Zone. *Proceedings of the Linnaean Society of New York,* 63–65:41–70.

Lack, D. 1947. *Darwin's Finches.* New York: Harper and Brothers.

Marquis, R. J., and C. J. Whelan. 1994. Insectivorous birds increase growth of white oak through consumption of leaf-chewing insects. *Ecology,* 75:2007–2014.

Martin, A. C., H. S. Zim, and A. L. Nelson. 1951. *American Wildlife and Plants: A Guide to Wildlife Food Habits.* London: Constable.

McAtee, W. L. 1920. *Local Suppression of Agricultural Pests by Birds.* Annual Report of the Smithsonian Institution.

Morse, D. H. 1968. The use of tools by Brown-headed Nuthatches. *Wilson Bulletin,* 80:220–224.

Rand, A. L. 1953. Factors affecting feeding rates of anis. *Auk,* 70:26–30.

Ray, T. S., and C. C. Andrews. 1980. Ant butterflies: butterflies that follow army ants to feed on antbird droppings. *Science,* 210:1147–1148.

Root, R. B. 1967. The niche exploitation pattern of the blue-gray gnatcatcher. *Ecological Monographs,* 37:317–350.

Schneirla, T. C. 1956. The army ants. *Annual Report of the Board of Regents of the Smithsonian Institution for 1955,* pp. 379–406.

Schorger, A. W. 1973. *The Passenger Pigeon.* Norman: University of Oklahoma Press.

Silloway, P. M. 1907. Stray notes from the Flathead Woods. *Condor,* 9:53–54.

Skead, C. J. 1951. Notes on honeyguides in southeastern Cape Province, South Africa. *Auk,* 68:52–62.

Terborgh, J. 1992. *Diversity and the Tropical Rain Forest.* New York: Scientific American Library.

Van Tyne, J. 1951. A Cardinal's, *Richmondena cardinalis,* choice of food for adult and for young. *Auk,* 68:110.

Willis, E. O., and Y. Oniki. 1978. Birds and army ants. *Annual Review of Ecology and Systematics,* 9:243–263.

## 4. The Bugs Fight Back

Annandale, N. 1900. Observations on the habits and natural surroundings of insects made during the "Skeat" expedition to the Malay Peninsula. *Proceedings of the Zoological Society of London,* 1900:837–869.

Bates, H. W. 1862. Contributions to an insect fauna of the Amazon Valley, Lepidoptera: Heliconidae. *Transactions of the Linnaean Society, Zoology,* 23:95–566.

Beehler, B. M., T. K. Pratt, and D. A. Zimmerman. 1986. *Birds of New Guinea.* Princeton: Princeton University Press.

Blest, A. D. 1957. The function of eyespot patterns in the Lepidoptera. *Behaviour,* 11:209–256.

Blum, M. S., D. W. Whitman, R. F. Severson, and R. F. Arrendale. 1987. Herbivores and toxic plants: evolution of a menu of options for processing allelochemicals. *Insect Science and Its Application,* 8:459–463.

Breed, M. D., G. E. Robinson, and R. E. Page, Jr. 1990. Division of labor during honey bee colony defense. *Behavioral Ecology and Sociobiology,* 27:395–401.

Bristowe, W. S. 1958. *The World of Spiders.* London: Collins.

Brower, J. V. Z. 1958. Experimental studies of mimicry in some North American butterflies. Part I. The monarch, *Danaus plexippus,* and viceroy, *Limenitis archippus archippus. Evolution,* 12:32–47.

Brower, L. P. 1969. Ecological chemistry. *Scientific American,* 220:22–30.

Cott, H. B. 1957. *Adaptive Coloration in Animals.* London: Methuen.

Curio, E. 1966. Wie Insekten ihre Feinde abwehren. (How insects ward off their enemies). *Naturwissenschaft und Medizin,* 11:3–22.

———1970. Validity of the selective coefficient of a behaviour trait in hawkmoth larvae. *Nature,* 222:382.

Dumbacher, J. P., B. M. Beehler, T. F. Spande, H. M. Garraffo, and J. W. Daly. 1992. Homobatrachotoxin in the genus *Pitohui:* chemical defense in birds? *Science,* 258:799–801.

Eisner, T. 1972. Chemical ecology: on arthropods and how they live as chemists. *Verhandlungsbericht der Deutschen Zoologischen Gesellschaft, 65. Jahresversammlung,* pp. 123–137.

Evans, D. L., and G. P. Waldbauer. 1982. Behavior of adult and naive birds when presented with a bumblebee and its mimic. *Zeitschrift für Tierpsychologie,* 59:247–259.

Hingston, R. W. G. 1932. *A Naturalist in the Guiana Forest.* New York: Longmans, Green.

Kettlewell, H. B. D. 1959. Darwin's missing evidence. *Scientific American,* 200:48–53.

Müller, F. 1879. *Ituna* and *Thyridis;* a remarkable case of mimicry in butterflies. Trans. R. Meldola. *Proceedings of the Entomological Society of London,* 27:xx–xxix.

Newnham, A. 1924. The detailed resemblance of an Indian lepidopterous larva to the excrement of a bird. A similar result obtained in an entirely different way by a spider. *Transactions of the Entomological Society of London,* 1924:xc–xciv.

Smith, S. M. 1975. Innate recognition of coral snake pattern by a possible avian predator. *Science,* 187:759–761.

———1977. Coral-snake pattern recognition and stimulus generalisation by naive great kiskadees. (Aves: Tyranidae). *Nature,* 265:535–536.

Sternburg, J. G., G. P. Waldbauer, and M. R. Jeffords. 1977. Batesian mimicry: selective advantage of color pattern. *Science,* 195:681–683.

Stowe, M. K., J. H. Tumlinson, and R. R. Heath. 1987. Chemical mimicry: bolas spiders emit components of moth prey species sex pheromones. *Science,* 236:964–967.

Waldbauer, G. P. 1988. Aposematism and Batesian mimicry. In M. K. Hecht, B. Wallace, and G. T. Prance, eds., *Evolutionary Biology,* 22:227–259.

———1988. Asynchrony between Batesian mimics and their models. In L. P. Brower, ed., *Mimicry and the Evolutionary Process.* Chicago: University of Chicago Press.

Wickler, W. 1968. *Mimicry in Plants and Animals.* Trans. R. D. Martin. New York: McGraw-Hill.

Winston, M. L. 1992. *Killer Bees: The Africanized Honey Bee in the Americas.* Cambridge, Mass.: Harvard University Press.

Yosef, R. and D. Whitman. 1993. An imperfect defense. *Living Bird,* Autumn 1993:27–29.

## 5. Bugs That Eat Birds

Askew, R. R. 1971. *Parasitic Insects.* New York: American Elsevier.

Atyeo, W. T., and R. M. Windingstad. 1979. Feather mites of the greater sandhill crane. *Journal of Parasitology,* 65:650–658.

Barker, S. C. 1994. Phylogeny and classification, origins, and evolution of host association of lice. *International Journal for Parasitology,* 24:1285–1291.

Bates, H. W. 1892. *The Naturalist on the River Amazons.* London: John Murray.

Bennett, G. F. 1961. On three species of Hippoboscidae (Diptera) on birds in Ontario. *Canadian Journal of Zoology,* 39:379–406.

Bray, O. E., J. J. Kennelly, and J. L. Guarino. 1975. Fertility of eggs produced on territories of vasectomized Red-winged Blackbirds. *Wilson Bulletin,* 87:187–195.

Bristowe, W. S. 1958. *The World of Spiders.* London: Collins.

Brown, C. R., and M. B. Brown. 1986. Ectoparasitism as a cost of coloniality in Cliff Swallows (*Hirundo pyrrhonota*). *Ecology,* 67:1206–1218.

Clay, T. 1949. Piercing mouth-parts in the biting lice (Mallophaga). *Nature,* 164:617–619.

———1949. Some problems in the evolution of a group of ectoparasites. *Evolution,* 3:279–299.

Comstock, J. H. 1950. *An Introduction to Entomology,* 9th ed. Ithaca, N.Y.: Comstock Publishing Co.

Cowan, F. 1865. *Curious Facts in the History of Insects.* Philadelphia: J. B. Lippincott.

Division of Entomology, Commonwealth Scientific and Industrial Research Organization of Australia. *The Insects of Australia.* Ithaca: Cornell University Press.

Dogiel, V. A. 1964. *General Parasitology.* Trans. Z. Kabata. London: Oliver and Boyd.

Gold, C. S., and D. L. Dahlsten. 1983. Effects of parasitic flies (*Protocalliphora* spp.) on nestlings of mountain and chestnut-backed chickadees. *Wilson Bulletin,* 95:560–572.

Green, R. G., C. A. Evans, and C. L. Larson. 1943. A ten-year population study of the rabbit tick *Haemaphysalis leporis-palustris. American Journal of Hygiene,* 38:260–281.

Halford, F. J. 1954. *Nine Doctors and God.* Honolulu: University of Hawaii Press.

Hamilton, W. D., and M. Zuk. 1982. Heritable true fitness and bright birds: a role for parasites? *Science,* 218:384–387.

Harrison, H. H. 1975. *A Field Guide to the Birds' Nests.* Boston: Houghton Mifflin.

Henshaw, H. W. 1902. *Birds of the Hawaiian Islands, Being a Complete List of the Birds of the Hawaiian Possessions with Notes on Their Habits.* Honolulu: Thomas G. Thrum.

Humphries, D. A. 1969. Behavioural aspects of the ecology of the sand-martin flea *Ceratophyllus styx jordani* Smit (Siphonaptera). *Parasitology,* 59:311–334.

Jellison, W. L. 1940. The burrowing owl as a host to the argasid tick, *Ornithodorus parkeri. U.S. Public Health Service, Public Health Reports,* 55 (part 1):206–208.

Johnson, L. L., and M. Boyce. 1991. Female choice of males with low parasite loads in sage grouse. In J. E. Loye and M. Zuk, eds., *Bird-Parasite Interactions.* Oxford: Oxford University Press.

Keirans, J. E. 1975. A review of the phoretic relationship between Mallophaga

(Phthiraptera: Insecta) and Hippoboscidae (Diptera: Insecta). *Journal of Medical Entomology,* 12:71–76.

Lofgren, C. S., W. A. Banks, and B. M. Glancey. 1975. Biology and control of imported fire ants. *Annual Review of Entomology,* 20:1–30.

Malcomson, R. O. 1960. Mallophaga from birds of North America. *Wilson Bulletin,* 72:182–197.

Marshall, A. G. 1981. *The Ecology of Ectoparasitic Insects.* London: Academic Press.

Merian, M. S. 1705. *Metamorphosis Insectorum Surinamensium.* Reprinted in 1982 with an English translation by P. A. van der Laan. Amsterdam: Jos. Fontaine/De Walburg Pers.

Mockford, E. L. 1967. Some Psocoptera from plumage of birds. *Proceedings of the Entomological Society of Washington,* 69:307–309.

Møller, A. P. 1991. Parasites, sexual ornaments, and mate choice in the barn swallow. In J. E. Loye and M. Zuk, eds., *Bird-Parasite Interactions.* Oxford: Oxford University Press.

Morris, C. D. 1988. Eastern equine encephalitis. In T. P. Monath, ed., *The Arboviruses: Epidemiology and Ecology.* Boca Raton: Chemical Rubber Company.

Nutting, W. B. 1976. Hair follicle mites (Acari) of man. *International Journal of Dermatology,* 15:79–98.

Packard, A. S. 1887. On the systematic position of the Mallophaga. *Proceedings of the American Philosophical Society,* 24:264–272.

Parikh, G. C., Z. D. Colburn, and D. R. Larson. 1969. Eastern equine encephalitis outbreak on a South Dakota pheasant farm. *Bacteriological Proceedings of the American Society for Microbiology,* 1969:159.

Perez, T. M., and W. T. Atyeo. 1984. Site selection of the feather and quill mites of Mexican parrots. In D. A. Griffiths and C. E. Bowman, eds., *Acarology* VI, vol. 1, pp. 563–570.

Rogers, C. A., R. J. Robertson, and B. J. Stutchbury. 1991. Patterns and effects of parasitism by *Protocalliphora sialia* on tree swallow nestlings. In J.E. Loye and M. Zuk, eds., *Bird-Parasite Interactions.* Oxford: Oxford University Press.

Rothschild, M., and T. Clay. 1952. *Fleas, Flukes, and Cuckoos.* London: Collins.

Savage, T. S. 1847. On the habits of the "drivers" or visiting ants of West Africa. *Transactions of the Entomological Society of London.* 5:1–15.

Savory, T. H. 1928. *The Biology of Spiders.* New York: MacMillan.

Schneirla, T. C. 1956. The army ants. *Annual Report of the Board of Regents of the Smithsonian Institution for 1955,* pp. 379–406.

Schnierla, T. C., ed. 1971. *Army Ants: A Study in Social Organization.* San Francisco: W. H. Freeman.

Shear, W. A. 1986. *Spiders.* Stanford: Stanford University Press.

Simmons, K. E. L. 1957. The taxonomic significance of the head-scratching methods of birds. *The Ibis*, 99:178–181.

Spurrier, M. F., M. S. Boyce, and B. F. J. Manly. 1991. Effects of parasites on mate choice by captive sage grouse. In J. E. Loye and M. Zuk, eds., *Bird-Parasite Interactions*. Oxford: Oxford University Press.

Stresemann, E. 1950. Birds collected during Capt. James Cook's last expedition (1776–1780). *Auk*, 67:66–68.

Travis, B. V. 1938. The fire ant (*Solenopsis* spp.) as a pest of quail. *Journal of Economic Entomology*, 31:649–652.

Usinger, R. L. 1966. *Monograph of Cimicidae (Hemiptera Heteroptera)*. Thomas Say Foundation, Entomological Society of America, vol. 7, pp. 1–585.

Viets, K. 1939. Eine Merkwürdige, neue, in Tiefsee-echiniden schmarotzende Halacaridengattung und-Art [Acari]. *Zeitschrift für Parasitenkunde*, 10:210–220.

Warner, R. E. 1968. The role of introduced diseases in the extinction of the endemic Hawaiian avifauna. *Condor*, 70:101–120.

Wooley, T. A. 1988. *Acarology*. New York: John Wiley and Sons.

## 6. The Birds Fight Back

Cade, T. J. 1973. Sun-bathing as a thermoregulatory aid in birds. *Condor*, 75:106–108.

Campbell, B., and E. Lack, eds. 1985. *A Dictionary of Birds*. Vermillion, S.D.: Buteo Books.

Clark, L. 1991. The nest protection hypothesis: the adaptive use of secondary compounds by European starlings. In J. E. Loye and M. Zuk, eds., *Bird-Parasite Interactions*. Oxford: Oxford University Press.

Clark, L. and J. R. Mason. 1985. Use of nest material as insecticidal and anti-pathogenic agents by the European starling. *Oecologia* (Berlin), 67:169–176.

———1987. Olfactory discrimination of plant volatiles by the European Starling. *Animal Behavior*, 35:227–235.

Clayton, D. H. 1991. Coevolution of avian grooming and ectoparasite avoidance. In J. E. Loye and M. Zuk, eds., *Bird-Parasite Interactions*. Oxford: Oxford University Press.

Clayton, D. H., and J. G. Vernon. 1993. Common grackle anting with lime fruit and its effect on ectoparasites. *Auk*, 110:951–952.

Condry, W. 1947. Behaviour of young carrion crow with ants. *British Birds*, 40:114–115.

Groff, E. M., and H. Brackbill. 1946. Purple grackles "anting" with walnut juice. *Auk*, 63:246–247.

Janzen, D. J. 1969. Birds and the ant × acacia interaction in Central America, with notes on birds and other myrmecophytes. *Condor,* 71:240–256.

Joyce, F. J. 1993. Nesting success of rufous-naped wrens *(Campylorhynchus rufinucha)* is greater near wasp nests. *Behavioral Ecology and Sociobiology,* 32:71–77.

Kelso, L., and M. M. Nice. 1963. A Russian contribution to anting and feather mites. *Wilson Bulletin,* 75:23–26.

Kennedy, R. J. 1969. Sunbathing behaviour of birds. *British Birds,* 62:249–258.

Ricklefs, R. E. 1979. *Ecology,* 2nd. ed. New York: Chiron Press.

Robinson, S. K. 1985. Coloniality in the yellow-rumped cacique as a defense against nest predators. *Auk,* 102:506–519.

Rothschild, M., and J. T. Clay. 1952. *Fleas, Flukes, and Cuckoos.* London: Collins.

Savage, T. S. 1847. On the habits of the "drivers" or visiting ants of West Africa. *Transactions of the Entomological Society of London,* 5:1–15.

Saxena, R. 1989. Insecticides from neem. In J. T. Aranson, B. J. R. Philogene and P. Morand, eds., *Insecticides of Plant Origin.* American Chemical Society Symposium Series, vol. 387, pp. 110–135.

Saxena, R. C., G. P. Waldbauer, N. J. Liquido, and B. C. Puma. 1981. Effects of neem seed oil on the rice leaf folder, *Cnaphalocrocis medinalis.* In H. Schmutterer, K. R. S. Ascher, and H. Rembold, eds., *Natural Pesticides from the Neem Tree,* Proceedings of the First International Neem Conference, pp. 189–203.

Sengupta, S. 1981. Adaptive significance of the use of margosa leaves in nests of house sparrows, *Passer domesticus. Emu,* 81:114–115.

Simmons, K. E. L. 1966. Anting and the problem of self-stimulation. *Journal of Zoology* (London), 149:145–162.

Skutch, A. F. 1945. The most hospitable tree. *The Scientific Monthly,* 60:5–17.

Smith, N. G. 1968. The advantage of being parasitized. *Nature,* 219:690–694.

———1980. Some evolutionary, ecological, and behavioural correlates of communal nesting by birds with wasps or bees. *Proceedings of the 17th International Ornithological Congress (Berlin, 1978),* vol. 2, pp. 1199–1205.

Stoddard, H. L. 1931. *The Bobwhite Quail.* New York: Charles Scribner's Sons.

Tinbergen, N. 1968. *Curious Naturalists.* Garden City, N.Y.: Doubleday.

Waldbauer, G. 1996. *Insects through the Seasons.* Cambridge, Mass.: Harvard University Press.

Whitaker, L. M. 1957. A résumé of anting, with particular reference to a captive orchard oriole. *Wilson Bulletin,* 69:195–262.

Wilson, E. O. 1971. *The Insect Societies.* Cambridge, Mass.: Harvard University Press.

Wimberger, P. H. 1984. The use of green plant material in bird nests to avoid ectoparasites. *Auk,* 101:615–618.

Wunderle, J. M., Jr., and K. H. Pollock. 1985. The bananaquit-wasp nesting association and a random choice model. *Ornithological Monographs,* 36:595–603.

## 7. Bugs That Eat People

Akre, R .D., A. Greene, J. F. MacDonald, P. J. Landolt, and H. G. Davis. 1981. *The Yellowjackets of America North of Mexico.* U.S. Department of Agriculture, Agricultural Handbook no. 552.

Anderson, J. F., and L. A. Magnarelli. 1994. Lyme disease: a tick-associated disease originally described in Europe, but named after a town in Connecticut. *American Entomologist,* 40:217–227.

Borg, A., and W. R. Horsfall. 1953. Eggs of floodwater mosquitoes. II. Hatching stimulus. *Annals of the Entomological Society of America,* 47:355–366.

Carson, R. 1962. *Silent Spring.* New York: Houghton Mifflin.

Cole, A. C. 1968. *Pogonomyrmex Harvester Ants: A Study of the Genus in North America.* Knoxville: University of Tennessee Press.

Comstock, J. H. 1950. *An Introduction to Entomology,* 9th ed. Ithaca, N.Y.: Comstock Publishing Co.

Costello, D. F. 1968. *The World of the Ant.* Philadelphia: J. B. Lippincott.

Flint, M. L., and R. van den Bosch. 1981. *Introduction to Integrated Pest Management.* New York: Plenum Press.

Foote, R. H., and H. D. Pratt. 1953. *The Culicoides of the Eastern United States (Diptera, Heleidae).* U.S. Public Health Service Monograph no. 18.

Hamilton, W. D., and M. Zuk. 1982. Heritable true fitness and bright birds: a role for parasites? *Science,* 218:384–387.

Hickey, J. L., and D. W. Anderson. 1968. Chlorinated hydrocarbons and eggshell changes in raptorial and fish-eating birds. *Science,* 162:271–273.

Holland, W. J. 1903. *The Moth Book.* Garden City, N.Y.: Doubleday, Page.

Hölldobler, B., and E. O. Wilson. 1990. *The Ants.* Cambridge, Mass.: Harvard University Press.

Horsfall, W. R. 1955. *Mosquitoes: Their Bionomics and Relation to Disease.* New York: Ronald Press.

Horsfall, W. R., H. W. Fowler, Jr., L .J. Moretti, and J. R. Larsen. 1973. *Bionomics and Embryology of the Inland Floodwater Mosquito,* Aedes vexans. Urbana: University of Illinois Press.

Hubbell, S. 1993. *Broadsides from the Other Orders.* New York: Random House.

James, M. T., and R. F. Harwood. 1969. *Medical Entomology.* London: MacMillan and Collier-MacMillan.

Johnson, L. L., and M. S. Boyce. 1991. Female choice of males with low parasite loads in sage grouse. In J. E. Loye and M. Zuk, eds., *Bird-Parasite Interactions.* Oxford: Oxford University Press.

Knab, F. 1906. The swarming of *Culex pipiens.* Psyche, 13:123–133.

McDaniel, B. 1979. *How to Know the Ticks and Mites.* Dubuque: William C. Brown.

Metcalf, C. L. 1932. *Black Flies and Other Biting Flies of the Adirondacks.* New York State Museum Bulletin 289.

Metcalf, R .L., and R. A. Metcalf. 1993. *Destructive and Useful Insects,* 5th ed. New York: McGraw-Hill.

Møller, A. 1991. Parasites, sexual ornaments, and mate choice in the barn swallow. In J. E. Loye and M. Zuk, eds., *Bird-Parasite Interactions.* Oxford: Oxford University Press.

Neal, T. J., and H. C. Barnett. 1961. The life cycles of the scrub typhus chigger mite, *Trombicula akamushi. Annals of the Entomological Society of America,* 54:196–203.

Pechuman, L. L., D. W. Webb, and H. J. Teskey. 1983. *The Diptera, or True Flies of Illinois. 1. Tabanidae.* Illinois Natural History Survey Bulletin 33.

Peterson, R. T. 1980. *A Field Guide to the Birds,* 4th ed. Boston: Houghton Mifflin.

Roberts, F. H. S. 1970. *Australian Ticks.* Melbourne: Commonwealth Scientific and Industrial Research Organization.

Spurrier, M. F., M. S. Boyce, and B. F. J. Manly. 1991. Effects of parasites on mate choice by captive sage grouse. In J. E. Loye and M. Zuk, eds., *Bird-Parasite Interactions.* Oxford: Oxford University Press.

Waldbauer, G. P. 1962. The mouthparts of female *Psorophora ciliata* (Diptera: Culicidae) with a new interpretation of the functions of the labral muscles. *Journal of Morphology,* 11:201–216.

Wilson, E. O. 1993. *The Diversity of Life.* Cambridge, Mass.: Harvard University Press.

Woodwell, G. M., C. F. Wurster, Jr., and P. O. Isaacson. 1967. DDT residues in an east coast estuary: a case of biological concentration of a persistent insecticide. *Science,* 156:821–824.

Wooley, T. A. 1988. *Acarology.* New York: John Wiley and Sons.

Wurster, C. F., Jr., D. H. Wurster, and W. N. Strickland. 1965. Bird mortality after spraying for Dutch elm disease with DDT. *Science,* 148:90–91.

## 8. People Fight Back

Busvine, J. R. 1976. *Insects, Hygiene, and History.* London: The Athlone Press.

Hansen, J. A. 1990. Presidential address: our race to the future. *Journal of the American Mosquito Control Association,* 6:361–365.

Headlee, T. J. 1945. *The Mosquitoes of New Jersey and Their Control.* New Brunswick, N.J.: Rutgers University Press.
Horsfall, W. R. 1962. *Medical Entomology.* New York: Ronald Press.
Kirby, W., and W. Spence. 1856. An *Introduction to Entomology,* 7th ed. London: Longman, Brown, Green, and Longmans.
Lehane, B. 1969. *The Compleat Flea.* New York: Viking Press.
Metcalf, R. L. 1994. Insecticides in pest management. In R. L. Metcalf and W. H. Luckman, eds., *Introduction to Insect Pest Management.* New York: John Wiley and Sons.
Oaks, S. C., Jr., V. S. Mitchell, G. W. Pearson, and C. C. J. Carpenter, eds. 1991. *Malaria: Obstacles and Opportunities.* Washington, D.C.: National Academy Press.
Rothschild, M. 1965. Fleas. *Scientific American,* 213:44–53.
———1965. The rabbit flea and hormones. *Endeavour,* 24:162–168.
Steingraber, E. 1957. *Antique Jewelry.* New York: Frederick A. Praeger.
Topsell, E. 1658. *The History of Four-footed Beasts and Serpents and Insects.* Reprinted in 1967 by DaCapo Press, New York.

## 9. A Brief Guide to the Insects

Borror, D. J., and R. E. White. 1970. *A Field Guide to the Insects of America North of Mexico.* Boston: Houghton Mifflin.
Borror, D. J., C. A. Triplehorn, and N. F. Johnson. 1987. *An Introduction to the Study of Insects,* 6th ed. Fort Worth: Harcourt Brace Jovanovich College Publishing.
Gill, F. B. 1990. *Ornithology,* 2nd ed. New York: W. H. Freeman.
Glassberg, J. 1993. *Butterflies through Binoculars.* New York: Oxford University Press.

## 10. Disappearing Diversity

Bent, A. C. 1958. *Life Histories of North American Blackbirds, Orioles, Tanagers, and Allies.* U. S. National Museum Bulletin 211.
Bohlen, D. 1989. *The Birds of Illinois.* Bloomington: Indiana University Press.
Campbell, R. W., T. R. Torgersen, and N. Srivastava. 1983. A suggested role for predaceous birds and ants in the population dynamics of the western spruce budworm. *Forest Science,* 29:779–790.
Collar, N. J., and A. T. Juniper. 1992. Dimensions and causes of the parrot conservation crisis. In S. R. Beissinger and N. F. R. Snyder, eds., *New World Parrots in Crisis.* Washington, D.C.: Smithsonian Institution Press.

Division of Endangered Species. 1996. *Endangered and Threatened Wildlife and Plants.* 50CFR 17.11 and 17.12. Washington, D.C.: U.S. Fish and Wildlife Service.

Ehrlich, P. R., D. S. Dobkin, and E. Wheye. 1992. *Birds in Jeopardy.* Stanford: Stanford University Press.

Hayden, T. 1996. *The Lost Gospel of the Earth.* San Francisco: Sierra Club Books.

Hendrickson, R. 1974. *Lewd Food.* Radnor, Penn.: Chilton Book Company.

Holmes, R.T ., T. W. Sherry, and F. W. Sturges. 1986. Bird community dynamics in a temperate deciduous forest: long term trends at Hubbard Brook. *Ecological Monographs,* 50:201–220.

Klein, B. C. 1989. Effects of forest fragmentation on dung and carrion beetle communities in central Amazonia. *Ecology,* 70:1715–1725.

Marquis, R. J., and C. J. Whelan. 1994. Insectivorous birds increase growth of white oak through consumption of leaf-chewing insects. *Ecology,* 75:2007–2014.

Matthiessen, P. 1987. *Wildlife in America.* New York: Viking Penguin.

Moors, P. J., ed. 1985. *Conservation of Island Birds.* Cambridge, Eng.: International Council for Bird Preservation.

Peters, C. M., A. H. Gentry, and R. O. Mendelsohn. 1989. Valuation of an Amazonian rainforest. *Nature,* 339:655–656.

Peterson, R. T. 1974. Birds. In *Proceedings of the Symposium on Endangered and Threatened Species of North America.* Washington, D.C.

Primack, R. B. 1993. *Essentials of Conservation Biology.* Sunderland, Mass.: Sinauer Associates.

Robinson, S. K., S. I. Rothstein, M. C. Brittingham, L. J. Petit, and J. A. Grzybowski. 1995. Ecology and behavior of cowbirds and their impact on host populations. In T. E. Martin and D. M. Finch, eds., *Ecology and Management of Neotropical Migrant Birds.* Oxford: Oxford University Press.

Schneider, S. H,. and L. Morton. 1981. *The Primordial Bond.* New York: Plenum Press.

Smith, T. B., R. K. Wayne, D. J. Girman, and M. W. Bruford. 1997. A role for ecotones in generating rainforest biodiversity. *Science,* 276:1855–1857.

Shelford, V. E. 1963. *The Ecology of North America.* Urbana: University of Illinois Press.

Torgersen, T. R. 1997. The bane of budworms. *Birder's World,* 11:42–45.

Wilson, E. O. 1984. *Biophilia.* Cambridge, Mass.: Harvard University Press.

———1992. *The Diversity of Life.* New York: W. W. Norton.

# Acknowledgments

I am greatly indebted to the many friends and colleagues who helped me by giving generously of their time and knowledge: John Anderson, May Berenbaum, Sam Beshers, Dan Blake, John Bouseman, Robert Braun, Lincoln Brower, Dale Clayton, Susan Fahrbach, Mary Flint, Judy Hansen, William Horsfall, Gail Kampmeier, Howard Kilpatrick, Rabbi Norman Klein, Jack Kuehn, Robert Lewis, the late Ellis MacLeod, Robert Marquis, Michael Martin, Robert Novak, Jimmy Olson, James Rising, Hugh Robertson, Gene Robinson, Scott Robinson, James Sternburg, Torolf Torgerson, and Donald Webb.

My wife, Stephanie Waldbauer, read and criticized the entire manuscript. Special thanks are due Jim Nardi, who did the drawings, and Jim Sternburg, who took most of the color photographs. Other color photographs came from Photo Researchers, Inc. (1: © Jeff Lepore; 8: © Ken M. Johns; 9: © John Mitchell; 38: © Kenneth H. Thomas) and from Daybreak Imagery (3, 5, 20, 21, and 28: © Richard Day; 34: © Todd Fink). Dorothy Nadarski patiently typed and retyped the manuscript. The book benefited greatly from the constant encouragement of Michael Fisher and Ann Downer-Hazell, the careful editing of Nancy Clemente, and the design expertise of Marianne Perlak.

# Index

Page numbers in **boldface** refer to illustrations.

## A

## C

# D

# E

## F

## G



## H

## I



## J



## K

## L



## M

## N

# O



# P

## Q



## R

## S

## T

# U



# V



# W

## Y



## Z